Data-Driven Modeling of Cyber-Physical Systems using Side-Channel Analysis

Sujit Rokka Chhetri
Mohammad Abdullah Al Faruque

Data-Driven Modeling of Cyber-Physical Systems using Side-Channel Analysis

 Springer

Sujit Rokka Chhetri
Department of Electrical Engineering
and Computer Science
University of California
Irvine, CA, USA

Mohammad Abdullah Al Faruque
Department of Electrical Engineering
and Computer Science (EECS)
University of California
Irvine, CA, USA

ISBN 978-3-030-37964-3   ISBN 978-3-030-37962-9   (eBook)
https://doi.org/10.1007/978-3-030-37962-9

This Springer imprint is published by the registered company Springer Nature Switzerland AG.
The registered company address is: Gewerbestrasse 11, 6330 Cham, Switzerland

We dedicate this book to our family.

Preface

Cyber-physical systems (CPS) consist of a unique integration of discrete cyber-domain processes and continuous physical domain components. Current modeling approaches use extensive first-principle approaches to derive the various components of CPS. However, it is difficult to model some of the stochastic phenomenon (such as environmental variation, physical process variation, etc.) of CPS using the first-principle approach. Hence, in this book we have explored a data-driven modeling approach for CPS and present various methodologies for modeling security and creating virtual replica or digital twin of the physical system. Furthermore, we will also present new algorithms to handle complex non-Euclidean data for modeling the CPS. More specifically, the book will present and exploration of the unintended emissions from the physical domain of the CPS to infer various cyber-domain states.

This book provides a new perspective on modeling the cyber-physical system using a data-driven approach. It covers the use of state-of-the-art machine learning and artificial intelligence algorithms for modeling various aspects of the CPS. It provides insight on how a data-driven modeling approach can be utilized to take advantage of the relation between the cyber and the physical domain of the CPS for aiding the first-principle approach in capturing the stochastic phenomenon affecting the CPS.

The books book provides a practical use case of the data-driven modeling approach for securing the CPS, presenting novel attack models, building and maintaining the digital twin of the physical system. Furthermore, it also provides novel data-driven algorithms to handle non-Euclidean data. In summary, this book presents a novel perspective for modeling the CPS.

The first-principle based approach for modeling the CPS is complex and time-consuming. Below we present three major reasons for proposing a new book in this area:

- Due to the advancement of machine learning and artificial intelligence algorithms, there has been a huge leap in performing data-driven modeling. However, to the best of our knowledge, there are no books covering the data-driven modeling of the CPS to aid in capturing the stochastic phenomenon affecting CPS.

- The book presents some practical application for securing the CPS as well as building the digital twin of the physical twin of CPS. The digital twin is expected to be one of the pillars for next generation of CPS. Hence, this book provides timely coverage of building and maintaining the digital twins of CPS.
- The book also provides novel algorithms for handling not just Euclidean data but also non-Euclidean data. These algorithms will thus demonstrate how the next generation of digital twins may be made more cognitive by allowing it to process and extract information from complex and higher dimensional data.

Some of the unique features of the book can be listed as follows:

- Only book covering the data-driven modeling of the CPS utilizing the unique relation between the cyber and the physical domain.
- Coverage of machine learning and artificial intelligence algorithms for data-driven modeling of the CPS.
- Practical use case of the data-driven modeling approach for security and building digital twin of the CPS.
- Well-structured and comprehensive book chapters covering the breadth and depth in data-driven modeling of CPS.

Irvine, CA, USA   Sujit Rokka Chhetri
Irvine, CA, USA   Mohammad Abdullah Al Faruque

Acknowledgments

We would like to thank all the current and past members of Advanced Integrated Cyber-Physical Systems (AICPS) lab at the University of California of Irvine for contributing to the research and content of this book. We are really grateful for your help and support.

Contents

Chapter 1
Introduction

1.1 Cyber-Physical System

Cyber-physical systems (CPS) consist of integration of computational components in the cyber-domain with the physical domain processes [1, 2]. The physical domain processes consist of actuators which are coordinated and controlled by the computational components via a communication network, where the computational processes are usually affected by the feedback provided by the sensors in the physical domain. In the cyber-domain, the computational and communication cores monitor and manipulate the discrete signals, whereas, in the physical domain, energy flows, which are mostly continuous domain signals, govern the physical dynamics of the system. Due to the juxtaposition of cross-layer components (*physical, network, control, system, operation*, etc.) and cross-domain components, CPS provides various technology solutions to multiple fields (*automotive, manufacturing, health care*, etc.) [3].

CPS consists of complex interaction between heterogeneous (different types of computation and communication platforms) and hybrid (discrete and continuous) components. Although integration of cyber and physical processes is covered by embedded systems, the networking and multi-domain interaction among embedded systems makes CPS system different and hence, challenging to design. Traditionally, there has been extensive research carried out for design and automation tools targeted at hardware and software co-design for embedded system design, and for multi-physics modeling and simulation of the control systems. However, due to the tight integration of cyber and physical domain unique challenges mostly related to integration are still being tackled. To this end, in this book we focus on modeling through utilization of cross-domain relationship of the CPS.

© Springer Nature Switzerland AG 2020
S. R. Chhetri, M. A. Al Faruque, *Data-Driven Modeling
of Cyber-Physical Systems Using Side-Channel Analysis*,
https://doi.org/10.1007/978-3-030-37962-9_1

1.2 Data-Driven Modeling

Traditionally, CPS modeling consists of various techniques depending on type of modeling (such as component-oriented, multi-agent-based, actor-oriented, event-based, or structural or behavioral based), based on the characteristics (such as domain-specific or multi-domain, timed or un-timed, discrete, continuous, or hybrid) or based on the modeling requirements (such as specification, analysis, verification, and validation [4]. Modeling consists of formal methodologies for designing and engineering CPS. Most of the existing modeling approaches are first-principle based, as they are derived after complex analysis by the domain experts [5]. There are various tools and resources for first-principle based CPS modeling. There are CPS modeling languages such as Architecture Analysis and Design Language(AADL) [6], Unified Modeling Language (UML) [7], Planning Domain Definition Language(PDDL) [8], CyPhyML [9], ESMol [10], etc. Moreover, there are various simulation tools such as Modelica, Simulink, LabView, Ptolemy [11], etc.

The first-principle based approach provides us with the CPS model which has formal and provable deterministic properties, and capable of enabling simulation and analysis for detecting design defects [12]. However, it has some limitations which arise due to the fact that they are poor at modeling non-deterministic or stochastic properties of CPS, which are introduced while physically realizing the model. These non-deterministic behaviors are introduced due to environmental variation, physical noise of the underlying platform, part failures, network delays, uncontrollable scheduling, etc. [13].

These limitations can be alleviated using data-driven modeling approaches. Data-driven modeling consists of utilizing the data about the specific system to estimate or infer the relationship between the system variables (for example, input and output variables) without requiring the domain expertise or detailed knowledge of the behavior of the system [14]. Although data-driven modeling required large amount of data, due to proliferation of sensors incorporated in CPS, high volume, velocity and variety of data can be acquired from operational CPS to aid in modeling the next generation of CPS. The main advantage of data-driven modeling approach is that it is able to utilize the side-channels (explained in next section) to infer various cross-domain system variables. Moreover, data-driven modeling methodology enjoys the advancement in the field of artificial intelligence, machine learning, data mining, etc., to build models capable of accurately estimating relationship between system variables. In this book, we explore various data-driven modeling methodologies to model the interaction between the cyber and the physical domain variables.

1.3   Side-Channel Analysis

Side-channel analysis consists of analyzing the analog emission from the physical domain of CPS to infer about the various system states (especially in the cyber-domain). Traditionally, side-channel was used to infer the cryptographic keys by utilizing the vulnerability in the physical implementation of the system which unintentionally leak the information [15] rather than using brute force or attacking theoretical weakness of the algorithms. In CPS, computational and communication cores, governed by cyber-processes, interact with physical domain sensors and actuators. This interaction is unique due to the fact that the information flow in the cyber-domain may manifests physically in the form of energy flows in the physical domain. These energy flows may be observed in the form of various analog emissions such as *vibration, acoustic, magnetic, power*, etc. These analog emissions may behave as side-channels and allow us to observe various cyber-domain state variables from the physical domain.

In this book, we present how a data-driven modeling can utilize the side-channels to infer about the relationship between cross-domain variables. The inference is performed to demonstrate vulnerability of the system to confidentiality breach, to build defense mechanism, to build digital twin of the CPS, and to perform non-euclidean data-driven modeling.

1.4   Book Sections

To cover various aspects of the data-driven modeling, we have divided the book into four parts. In Parts I and II we focus on data-driven modeling methodology for security (attack modeling and defense) of cyber-physical systems. In Part III we present data-driven modeling methodology for building the digital twin of cyber-physical systems. Finally, in Part IV, we present algorithms for performing non-euclidean data-driven modeling.

1.4.1   Part I: Data-Driven Attack Modeling

Attack modeling for cyber-physical system consists of defining various aspects such as intent of an adversary, outcome of the attack, target system, existing vulnerability of the system, medium of an attack, capability of an attacker, resource used by an attacker, and the cost of the attack. Researchers have extensively worked to formally characterize the attack models through various means [16] (such as description of consequences, separation of CPS security with the traditional information technology security, domain description, etc.). This characterization is done with the hope of paving the path for exhaustively exploring all the attack models. With the exploration of the attack models, the design and modeling of cyber-physical system can be influenced to strengthen the security. However, every year researchers have discovered new attack models that have not been considered earlier, which only proves the herculean nature of discovering the attack models.

In addition, the integration of the cyber and the physical domain of CPS makes it even more challenging to discover all possible attack models. Hence, in this part of the book, we focus on data-driven attack modeling aspect of the cyber-physical systems. Data-driven modeling approach allows us to analyze the information and the energy flows of the system to develop models quantify the information leakage from observable variables of the system. In the data-driven attack modeling paradigm, an attacker is assumed to treat the system as a black box. They can only observe the input and the output variables of the profiling system. This profiling system, which has similar built, is assumed to be accessible by the attacker. An attacker then uses statistical and probabilistic tools to collect and estimate function to infer the input and output relation during the training phase. Then they will utilize the estimated function to carry out the attack phase. In this part of the book, we will analyze various practical data-driven attack models of cyber-physical system.

1.4.2   Part II: Data-Driven Defense of Cyber-Physical Systems

In Part I of this book we present data-driven attack modeling. In Part II, we provide data-driven defense methodologies for securing the cyber-physical systems. Researchers have provided various formal ways for attack detection and defense in cyber-physical system [17, 18]. In this part, the focus is given to data-driven modeling approaches that utilize the unique relation between the physical and the cyber-domain. Moreover, physical emissions from the CPS that behave as side-channels are utilized to characterize the system. This characterization is done utilizing the data-driven modeling approaches.

To narrow down the scope, this part focuses on strengthening the confidentiality of the system, and detecting the cross-domain attacks. The cross-domain attacks considered in this book consists of attacks initiated in the cyber-domain and having consequences in the physical domain. Furthermore, we present a chapter where a data-driven security modeling tool is presented to provide system level security analysis.

1.4.3   Part III: Data-Driven Digital Twin Modeling

A *digital twin* is the virtual representation of a physical system (its *physical twin*) [19]. The concept of the *digital twin* was first used by NASA to describe a digital replica of physical systems in space maintained for diagnosis and prognosis. Digital twin consists of large historical context and performance data and utilizes the direct (through inbuilt sensors) and indirect (through latent variable analysis) sensing to provide the near real-time representation of the physi+ component, etc.,) which can provide the blueprint of the whole system [20]. The digital twin allows the user to monitor, simulate, optimize, and control the entire physical system in the virtual domain. Organizations like Siemens [21], General Electric [22], NASA [23], and the Air Force [24] are currently building *digital twins* of gas turbines, wind turbines, engines, and airplanes that allow them to manage the assets, optimize the system and fleets, and to monitor the system health and provide prognostics.

Digital twins can be modeled to accurately estimate the interaction between the cyber and the physical components by building data-driven models to observe the input (cyber-domain variables) and the output (side-channels emissions) variables of the system. Since the side-channel rely on the physical implementation of the CPS, they are able to capture various stochastic behaviors in CPS that arise due to the environmental variation, human interaction, process variation, etc. In this part, we present methodologies to estimate the digital twin using data-driven modeling methodology by observing the side-channels.

1.4.4   Part IV: Non-Euclidean Data-Driven Modeling of Cyber-Physical Systems

Data-driven modeling approaches utilize various algorithms from the field of machine learning and artificial intelligence. Over recent years, due to availability of large amount of data and decrease in price of computation, complex algorithms such as deep neural network have been able to perform data-driven modeling with high accuracy in task such as classification and regression. Most of the algorithms developed so far are, however, for euclidean space. Euclidean data are defined as set of points in euclidean space which satisfy certain relationship or are expressible in terms of distance and angle. They can be operated using various well known operations such as euclidean norm, dot product, etc. Some of the examples of these data are time-series data, image, etc.

However, in CPS, there are large amount of data which are non-euclidean in nature. These data consists of no common system of coordinates, no vector space structure, no description of property like shift in-variance in convolutional neural network for images to name few. Examples of non-euclidean data include social network data, graphs, manifolds, etc. In CPS, there are large amount of non-euclidean data, such as call-graphs, finite-state-machines, etc. The engineering design data mostly varies from domain to domain. In electronic design, it consists of high-level design descriptions, register transfer level descriptions in Verilog or VHDL, schematics, etc. In mechanical design, it consists of data regarding structural designs, modeling, and analysis of components, etc. Moreover, there is a wealth of data generated throughout the supply chain of engineering including computer-aided design (CAD) and computer-aided manufacturing (CAM) tools. In order to perform meaningful learning from these data, we need to utilize non-euclidean or graph learning algorithms that are able to extract, categorize, and label these sparse data. Hence, in this part, to aid data-driven modeling, novel algorithms capable of handling non-euclidean data are presented.

1.5   Summary

In summary, this book focuses on the following major challenges for data-driven modeling of cyber-physical systems.

- Modeling interaction between cyber and physical system.
- Modeling non-functional requirements of CPS.
- Modeling and updating digital twins of legacy CPS.
- Handling non-euclidean data.

To address the above mentioned challenges, this book presents the following solutions:

1. Novel data-driven modeling for exploring new attack models (Part I).
2. Cross-domain security models for integrity (Part II).
3. Internet of Things (IoT)-enabled living digital twin for legacy CPS (Part III).
4. Novel algorithms for handling non-euclidean data (Part IV).

References

1. Rajkumar, R., Lee, I., Sha, L., & Stankovic, J. (2010). Cyber-physical systems: the next computing revolution. In *Design Automation Conference* (pp. 731–736). Piscataway, NJ: IEEE.
2. Lee, E. A. (2008). Cyber physical systems: Design challenges. In *2008 11th IEEE International Symposium on Object Oriented Real-Time Distributed Computing (ISORC)*. Piscataway, NJ: IEEE.
3. Shi, J., Wan, J., Yan, H., & Suo, H. (2011). A survey of cyber-physical systems. In *2011 International Conference on Wireless Communications and Signal Processing (WCSP)* (pp 1–6). Piscataway, NJ: IEEE.
4. Graja, I., Kallel, S., Guermouche, N., Cheikhrouhou, S., & Hadj Kacem, A. (2018). A comprehensive survey on modeling of cyber-physical systems. *Concurrency and Computation: Practice and Experience* (p. e4850).
5. Lam, R. R., Horesh, L., Avron, H., & Willcox, K. E. (2017). Should you derive, or let the data drive? An optimization framework for hybrid first-principles data-driven modeling. arXiv:1711.04374.
6. Franca, R. B., Bodeveix, J.-P., Filali, M., Rolland, J.-F., Chemouil, D., & Thomas, D. (2007). The AADL behaviour annex–experiments and roadmap. In *Proceedings of the 12th IEEE International Conference on Engineering Complex Computer Systems (ICECCS 2007)* (pp. 377–382). Piscataway, NJ: IEEE.
7. Bock, C. (2006). SysML and UML 2 support for activity modeling. *Systems Engineering, 9*(2), 160–186.
8. Constructions Aeronautiques, Howe, A., Knoblock, C., McDermott, I. D., Ram, A., Veloso, M., et al. (1998). PDDL - The Planning Domain Definition Language.

9. Sztipanovits, J., Bapty, T., Neema, S., Howard, L., & Jackson, E. (2014). OpenMETA: A model-and component-based design tool chain for cyber-physical systems. In *From programs to systems. The systems perspective in computing* (pp. 235–248). Berlin: Springer.
10. Kottenstette, N., Karsai, G., & Sztipanovits, J. The ESMoL language and tools for high-confidence distributed control systems design. Part 1: Design language, modeling framework, and analysis. *ISIS, 10*, 109.
11. Buck, J. T., Ha, S., Lee, E. A., & Messerschmitt, D. G. (1994). Ptolemy: A framework for simulating and prototyping heterogeneous systems.
12. Sharma, A. B., Ivančić, F., Niculescu-Mizil, A., Chen, H., & Jiang, G. (2014). Modeling and analytics for cyber-physical systems in the age of big data. *ACM SIGMETRICS Performance Evaluation Review, 41*(4), 74–77.
13. Lee, E. (2015). The past, present and future of cyber-physical systems: A focus on models. *Sensors, 15*(3), 4837–4869.
14. Solomatine, D., See, L. M., & Abrahart, R.J. (2009). Data-driven modelling: Concepts, approaches and experiences. In *Practical hydroinformatics* (pp. 17–30). Berlin: Springer.
15. Standaert, F.-X., Malkin, T. G., & Yung, M. (2009). A unified framework for the analysis of side-channel key recovery attacks. In *Annual International Conference on the Theory and Applications of Cryptographic Techniques*. Berlin: Springer.
16. Cardenas, A., Amin, S., Sinopoli, B., Giani, A., Perrig, A., & Sastry, S. (2009). Challenges for securing cyber physical systems. In *Workshop on Future Directions in Cyber-Physical Systems Security*.
17. Pasqualetti, F., Dörfler, F., & Bullo, F. (2012). Attack detection and identification in cyber-physical systems–part II: Centralized and distributed monitor design. arXiv:1202.6049
18. Giraldo, J., Urbina, D., Cardenas, A., Valente, J., Faisal, M., Ruths, J., et al. (2018). A survey of physics-based attack detection in cyber-physical systems. *ACM Computing Surveys, 51*(4), 76.
19. Negri, E., Fumagalli, L., & Macchi, M. (2017). A review of the roles of digital twin in CPS-based production systems. *Procedia Manufacturing, 11*, 939–948.
20. Grieves, M., & Vickers, J. (2017). Digital twin: Mitigating unpredictable, undesirable emergent behavior in complex systems. In *Transdisciplinary Perspectives on Complex Systems*. Berlin: Springer.
21. Schmitt, R., Rose, S., et al. (2015). Advance: Digital Enterprise—on the way to industrie 4.0. https://goo.gl/LtL5oy.
22. General Electric Company. (2016). GE Digital Twin: Analytic engine for the digital power plant. https://goo.gl/bhQzRF.
23. Glaessgen, E., & Stargel, D. (2012). The digital twin paradigm for future NASA and US Air Force vehicles. In *Structures, Structural Dynamics and Materials Conference*.
24. Kraft, E. M. (2016). The Air Force Digital thread/digital twin-life cycle integration and use of computational and experimental knowledge. In *54th AIAA Aerospace Sciences Meeting*.

Part I
Data-Driven Attack Modeling

Chapter 2
Data-Driven Attack Modeling Using Acoustic Side-Channel

2.1 Introduction

Cyber-physical systems consist of the integration of computation, physical, and networking components. The synergy of these components results in a new form of vulnerabilities, which cannot be addressed by traditional security solutions designed for the individual components. In this chapter, the focus is put on cyber-physical additive manufacturing systems, where 3D objects are created one layer at a time. Fused deposition modeling (FDM) is one of the technologies used in additive manufacturing, where plastic or metal filaments, heated slightly above their melting point, are deposited to construct a 3D object. Several sectors, such as medical and aerospace, are increasingly adopting the use of these additive manufacturing systems [1, 2]. In addition, agencies like the U.S. Air Force, Navy, and NASA are also incorporating additive manufacturing into their manufacturing processes [2, 3]. This trend shows that new vulnerabilities, such as cross-domain attacks, can have a large economic impact on manufacturing industries utilizing additive manufacturing. Attackers who target cyber-physical manufacturing systems are often motivated by either industrial espionage of intellectual property (IP), alteration of data, or denial of process control [4, 5]. The world economy relies heavily on IP-based industries, which produce and protect their designs through IP rights. IP in cyber-physical manufacturing consists of *the internal and external structure of the object*, *the process parameters*, and *the machine-specific tuning parameters* [6]. To produce a 3D object, design information (which contains IP) is supplied to the manufacturing system in the form of G-code. G-code, a programming language, is primarily used in FDM to control the system components and parameters such as *speed, extrusion amount*, etc [7]. If these designs are stolen, they can be manipulated to harm the image of the company, or even worse, can cause the company to lose its IP (as it is stolen before production) [8]. Currently, IP theft mainly occurs through the cyber-domain (e.g., Operation Aurora, GhostNet) [9, 10], but IP information can also be

© Springer Nature Switzerland AG 2020   11
S. R. Chhetri, M. A. Al Faruque, *Data-Driven Modeling
of Cyber-Physical Systems Using Side-Channel Analysis*,
https://doi.org/10.1007/978-3-030-37962-9_2

leaked through the physical-domain (side-channels). A common example of this is to use side-channel information (e.g., timing data, acoustics, power dissipation, and electromagnetic emission) from devices performing the cryptographic computation to determine their secret keys [11]. In this chapter, a data-driven modeling approach is adopted to highlight the possibility of physical-to-cyber attacks on CPS, and motivate a general research interest in novel ways to minimize side-channel leakage during design time and manufacturing.

2.1.1   Research Challenges and Contributions

It is probably not possible to make a system completely secure. This is because many vulnerabilities are not known during design time. Hence, it is necessary to continue to investigate this field to identify novel threats. The contribution of the research presented in this chapter [12] to the security research in cyber-physical additive manufacturing systems are as follows:

- **Source of acoustic emission (Sect. 2.3)** where the source of acoustic emission in an FDM technique based 3D printer is analyzed, and various equations to better understand the working principles of the proposed attack model are presented.
- **Acoustic leakage analysis (Sect. 2.4)** where the side-channel leakage model is provided and leakage quantification is performed to understand the relation between the cyber-data and acoustic emission.
- **A data-driven acoustic side-channel attack model to breach confidentiality (Sect. 2.5)** which describes our novel attack methodology. It consists of an exploration of time and frequency domain features, learning algorithms trained to acquire specific information (axis of movement, speed of the nozzle) about the G-code, context-based post-processing, and algorithms used to reconstruct the G-code by reverse engineering.

2.2   Background and Related Work

A typical digital process chain in cyber-physical additive manufacturing systems is presented in Fig. 2.1. Designers start their design of 3D objects with 3D computer-aided design (CAD) modeling tools such as Sketchup [13] and the extended version of Photoshop [14]. Next, the CAD tool generates a standard stereolithography (STL) file for the manufacturing purpose. Computer-aided manufacturing (CAM) software is then required to slice the STL file into layer-by-layer description file (e.g., G-code, cube, etc.). Then, the layer description file is sent to the manufacturing system (e.g., 3D printer) for production [7].

In the physical-domain of the cyber-physical additive manufacturing system, components such as a stepper motor, fan, extruder, base plate, etc., carry out

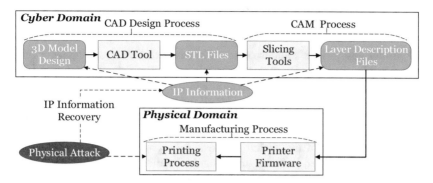

Fig. 2.1 Physical attack during printing process of cyber-physical additive manufacturing system

operations on the basis of information provided by the cyber-domain (G-code). In carrying out the operation, these physical components leak cyber-domain information (G-code) from the side-channels, such as acoustic and power, which may be used to steal IP by performing a physical-to-cyber attack. The issues regarding the theft of IP and the framework for preventing IP theft have been studied in [6, 15]. The study of attack in the process chain, starting from the 3D object design to its creation, along with a case study of cyber-attacks in the STL file, is presented in [16]. However, physical-domain attacks are not well studied by the existing works.

There are several publications utilizing the side-channel information to gather data related to the cyber-domain in other systems. [17] have used the acoustics emanated from the dot matrix printer while printing to recover the text it was sent to print. Authors in [18] have been able to decode the keys pressed in the Enigma machine by analyzing the sound made by the device while pressing the keys. However, these methodologies are not applicable to 3D printers since, unlike printed words on paper, a 3D printer's movement has infinite possibilities. Researchers from MIT have found that even the minor movement of physical devices can leak information about the cyber-domain. In [19], they have successfully retrieved digital audio being played by capturing the vibration of objects near a sound source by a high-speed camera. However, in a 3D printer, there are multiple sources of sound and vibration. Therefore, the task of analyzing sound for G-code reconstruction requires a completely new approach. Authors in [20] have considered using side-channel for providing security, but they have not demonstrated any methodology for using it to steal the IP. In summary, the literature is focused on retrieving the text being printed (either in a keyboard or dot matrix printer), analyzing acoustic emissions for observing mechanical degradation of the physical components in a manufacturing plant, etc. However, the possibility of using the acoustic emissions for the reconstruction of a 3D object has not been considered. Hence, in this chapter, a data-driven acoustic side-channel attack model to breach the confidentiality of cyber-physical additive manufacturing systems is presented.

2.3 Sources of Acoustic Emission

A 3D printer has various sources of acoustic emissions. A strong attacker may be able to describe the behavior of acoustic emission in the side-channel based on the various G-codes using a single physical model (using a single equation). However, this task is not trivial, as it has to consider all the sources of vibration in the system. Moreover, they may require the system design knowledge to model the mechanical and electrical model of the system, which may not readily be available. Instead, data-driven modeling of the system can easily be created by treating the system as a black box. However, to better understand why these data-driven models work, a preliminary analysis of individual sources of sound in a typical FDM technique based 3D printer is provided in this chapter. Based on this, in Sect. 2.4, leakage analysis is performed to understand why various information about G-code can be inferred from the acoustic emission.

2.3.1 System Description

State-of-the-art FDM based additive manufacturing systems consist of four to five stepper motors depending upon their structural design and number of filaments available for extrusion. Due to high torque/size ratio, hybrid two-phase stepper motors have been widely used in these 3D printers. However, these stepper motors are the major source of sound that enables the leakage of cyber-domain information from the side-channel. Hence, in the rest of the sections, the focus is given on describing various mathematical models of the stepper motor and analyzing how they aid in sound production.

The energy conversion steps for a stepper motor is shown in Fig. 2.2. Here, the electrical energy (current i) is first converted to an electric and magnetic field, which in turn guides the rotors. The electromagnetic field acting upon the various components produces force F_{em}, which causes them to vibrate and produce sound with power P. However, the electrical energy is controlled by the G-codes, which gives the printer instruction to control the dynamics of the system (such as nozzle speed, axis movement, etc.). Hence, the acoustics emitted by the printer eventually depends on the supplied G-code. Moreover, the stepper motor behaves like a traditional audio speaker due to its structure (see Fig. 2.2b).

2.3.2 Equation of Motion

To understand the production of mechanical energy, the equation of motion for a hybrid stepper motor can be analyzed as follows [21]:

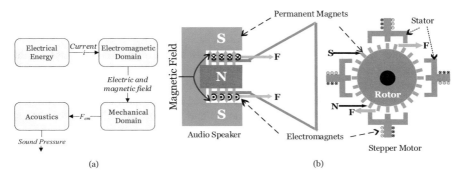

Fig. 2.2 Internal structure of the stepper motor. (**a**) Energy flow in stepper motor. (**b**) Comparison of stepper motor with audio speaker

$$J\frac{d^2\theta}{dt^2} + D\frac{d\theta}{dt} + p\Psi_m i_A sin(p\theta) + p\Psi_m i_B sin(p(\theta - \lambda)) = 0 \qquad (2.1)$$

where J is the moment of inertia of the rotor and the load combined, $J = J_M + J_L$, and D is the damping coefficient based on eddy current, air friction, hysteresis effects, etc. i_A and i_B are the current flowing through the two phases. ψ_m is the maximum stator flux linkage, p is the number of rotor pole pairs, λ is the angle between the two stator winding, and θ is the mechanical rotational angle. Equation 2.1 states that the inertia in the motor depends on the current being supplied to the two phases of the stator. This inertia determines the amount of vibration produced by the motor when it is rotating. Moreover, based on this equation the natural oscillation frequency of the rotor can be derived, which plays a major role in generating unique acoustics for the given stepper motor.

2.3.3   Natural Rotor Oscillation Frequency

The radiated sound power is higher when the stepper motor vibrates with the rotor's natural oscillation frequency. Using Eq. 2.1, the natural frequency of rotor oscillation can be calculated as follows [22]:

$$\omega_{np}^2 = \frac{2p^2\Psi_m I_o cos(\frac{p\lambda}{2})}{J} \qquad (2.2)$$

where I_o is the stationary current flowing in the two phases A and B. When the stepper motor is rotating with the harmonic frequency of the natural frequency such as $\ldots, \frac{\omega_{np}}{4}, \frac{\omega_{np}}{3}, \frac{\omega_{np}}{2}, 2\omega_{np}, 3\omega_{np}, 4\omega_{np}, \ldots$ the vibration is more prominent due to resonance. Given the fact that J is the inertia of load and the rotor combined, varying load in the stepper motor will change its natural rotor oscillation frequency as well.

2.3.4   Stator Natural Frequency

Natural rotor oscillation frequency corresponds to the dynamic response of the stepper motor. However, the stator itself has a natural frequency which depends on various parameters. One of the major parameters is vibration modes. Due to the prominence of the radial force acting on the stator, consideration may be given only to the circumferential radial vibration modes and the corresponding stator natural frequencies. The structure of the stator is complex and many attempts have been made to calculate the natural frequencies of the stator with various considerations, an example being single-ring type stator [23, 24]. Since the external structure connected to the stator also influences its mass and stiffness, the natural frequency of the stator with circumferential vibration mode m and axial vibration mode n of the frame may be calculated as follows [25]:

$$\omega^2_{stator\ np} \approx \frac{K_m^{(c)} + K_{mn}^{(f)}}{M_c + M_f} \tag{2.3}$$

where $K_m^{(c)}$ is the lumped stiffness of the stator core, $K_{mn}^{(f)}$ is the lumped stiffness of the frame, and M_c and M_f are the mass of the stator and the frame, respectively. Equation 2.3 has been derived by assuming that the lumped stiffness of the core and the frame are in parallel.

2.3.5   Source of Vibration

The main sources of vibration in stepper motors are electromagnetic, mechanical, and aerodynamic [26]. These vibrations help in radiating sound from the stepper motor stator surface and the frame to which the motor is connected. In this chapter, consideration is given to the electromagnetic and mechanical sources as they are the major sources of leakage.

2.3.5.1   Electromagnetic Source

The fundamental source of vibration in hybrid stepper motors is due to the fluctuation of electromagnetic force produced by the winding of the stator. The two types of vibration produced by the electromagnetic force are:

1. *Radial Stator Vibration*: In a hybrid stepper motor, both stator and the rotor are responsible for exciting the magnetic flux density in the air gap between the rotor and the stator. These magnetic fluxes contribute to generating the radial force. If $\sigma_{l,k}$ be the radial force at pole l for kth harmonic, then the total radial force acting on the stepper motor may be calculated as follows [27]:

Fig. 2.3 Outer mechanical
structure of a stepper motor

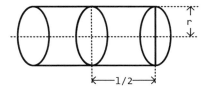

$$\sigma_{total} = \sum_{k=1}^{\infty} \sigma_{l,k} cos(k\omega t + \phi_{l,k}) \tag{2.4}$$

where $\phi_{l,k}$ is the phase angle of the radial force at pole l for k harmonic, $\omega = 2\pi f$, and f is the frequency determined by the stepping rate of the motor. This radial force acts on the stator and rotor surface and deforms its structure. This produces vibration and eventually sound in the stepper motor. When the radial force excites the harmonics of the natural frequencies of the stator/frame structure and the rotor oscillation, vibration is more prominent due to resonance.

Let us assume that the structure of the stator of the hybrid-stepper motor is cylindrical (as shown in Fig. 2.3). With this, we may express the total sound power radiated by the electric machine due to the varying radial force acting upon the stator as follows [23, 26]:

$$P = 2\rho c\pi^2 f^2 A_{rd}{}^2 (4\pi rl/2) I_{rel} = 4\rho c\pi^3 f^2 A_{rd}{}^2 rl I_{rel} \tag{2.5}$$

where P is the radiated sound power (W), ρ is the density of the medium (kg/m^3), c is the speed of the sound in the medium (m/s), f is the excitation frequency of the vibration with multiple harmonics (Hz), A_{rd} is the surface vibratory displacement (m), r is the radius of the cylindrical stator (m), l is the length of the stepper motor (m), and I_{rel} is the relative sound intensity. I_{rel} depends on the mode of stator vibration R, the radius, and the length-diameter ratio. Hence, stepper motors with different geometry and design in the 3D printer will emit different sound power.

2. *Torque Ripple*: Even though torque ripple is substantially reduced by using the micro-stepping for driving the stator windings, micro-stepping position ripple is still produced due to non-conformity to the ideal sine/cosine waves required for absolute removal of the torque ripple. However, the vibration produced by the torque ripple is less compared to the radial stator vibration.

2.3.5.2   Mechanical Source

The rotor and load connected to the stepper motor may also produce vibration and sound at various frequencies due to friction, rotor unbalance, shaft misalignment, loose stator lamination, etc. These vibrations produce a loud noise due to resonance.

In summary, the major source of acoustic emission are the stepper motors. However, any actuator (for example, DC motors used in cooling fans) that responds to the G-code is also capable of leaking information from the acoustic side-channel. A strong attacker will consider all these facts to make their attack model accurate. In the subsequent sections, a detailed analysis of the information that may be leaked through the acoustic emission is presented.

2.4 Acoustic Leakage Analysis

2.4.1 Side-Channel Leakage Model

Using an acoustic data acquisition device, an attacker may physically observe analog emissions $[o_1, o_2, o_3, \ldots, o_i]$, where o_i denotes the ith sample observed in the time domain. This is in fact the measurement of the acoustic power radiated by the stepper motor, as given in Eq. 2.5. Let O be a random variable denoting observable analog emissions, then side-channel leakage can be modeled as follows:

$$O = f_d(.) + N \tag{2.6}$$

where $f_d(.)$ represents a deterministic leakage function that may be modeled by an attacker. N represents a random variable denoting noise independent from $f_d(.)$ added to the side-channel. Leakage function $f_d(.)$ depends on the G-code instruction $[g_1, g_2, g_3, \ldots, g_j]$, where g_j is the jth instruction supplied to the 3D printer. The G-code may be denoted by a random variable G, then the leakage model may be re-written as follows:

$$O = f_d(G) + N \tag{2.7}$$

Since some of the G-code take longer time to execute than others, for each G-code instruction g_j, there will be k samples of analog emissions corresponding to it. The length k depends on the sampling frequency of the audio device used. The fundamental information contained in g_j are speeds in each axis $\{v_{x_j}, v_{y_j}, v_{z_j}\} \in \mathbb{R}_+$, presence or absence of axis movements $\{a_{x_j}, a_{y_j}, a_{z_j}\} \in \{0, 1\}$ with 0 representing absence of movement, positive or negative distance moved in each axis $\{d_{x_j}, d_{y_j}, d_{z_j}\} \in \mathbb{R}$, and extrusion amount $d_{e_j} \in \mathbb{R}$. Hence, Eq. 2.7 may be rewritten as follows:

$$o_i = f_d(v_{x_i}, v_{y_i}, v_{z_i}, a_{x_i}, a_{y_i}, a_{z_i}, d_{x_i}, d_{y_i}, d_{z_i}, d_{e_i}) + n_i \tag{2.8}$$

where the smaller alphabets represent the value of the random variables O, G, and N during ith time interval, with random variable G further partitioned into random variables $\{V_x, V_y, V_z, A_x, A_y, A_z, D_x, D_y, D_z, D_e\}$ representing speed, axis movement, and distance, respectively, with their corresponding values passed to the function $f_d(.)$ in Eq. 2.8. Using data-driven approach, various machine learning algorithms such as regression, classifications, etc., may be used to estimate a function, such that for the ith sample of observable emissions the corresponding G-code $\hat{g}_i = \hat{f}_d(o_i, \alpha) + n_i$ can be estimated, where $\hat{f}_d$ represents the estimated function and α represents the tuning parameter learned for the function. Due to the presence of multiple parameters, classification and regression machine learning algorithms may be used to estimate multiple functions to estimate individual parameters separately.

Assumption Given the acoustic leakage O, the frequency of the radiated sound varies according to the speed of the nozzle in the X and Y axes, respectively.   □

The radial force generated in Eq. 2.4 in each pole depends on the magnetic flux density, stator tooth width, and rotor cap thickness. The magnetic flux density depends on the current passing through each winding. To increase the angular speed of the stepper motor, the stepping rate is increased. From Eq. 2.4, it can be seen that this increases the frequency of the radial force acting on the stepper motor. From Eq. 2.5, it can also be seen that the radiated power increases with the excitation frequency of the vibration.

Assumption Given the acoustic leakage O, the power frequency spectrum of the radiated sound from the stepper motors X, Y, Z, and the one for the extruder are different.   □

The natural rotor oscillation frequency in Eq. 2.2 is inversely proportional to the moment of inertia of the load and the motor ($J_L + J_M$). The load moved by each stepper motors X, Y, Z, and E is different in state-of-the-art stepper motors. The natural frequency in Eq. 2.3 also depends on the mechanical structure of the frame to which the stepper motor is connected. Due to the mechanical structure of the 3D printers, stepper motors are placed in various locations and are connected to different frame structures. Therefore, the natural frequencies of the stepper motors vary according to the load and the frame to which they are attached. This means that the resonance can occur at different frequencies of the vibration for different stepper motors and the frame structure. This causes the power spectrum of the radiated sound to vary according to the source of the sound, i.e., the stator motor and the frame structure.

Assumption Given the acoustic leakage O, the intensity of the radiated power will vary according to the direction of the nozzle movement in different directions from the audio device.   □

According to the inverse square law, the intensity of the sound decreases drastically with the square of the distance from the sound source. If P is the power of the sound source and r be the distance from the sound source, then:

$$I = \frac{P}{4\pi r^2} \tag{2.9}$$

Hence, for analyzing the direction of movement, the intensity of the sound radiated by each motor and frame structure may be measured.

Assumption The direction of movement in Z axis during printing is always in either a positive or negative direction. $\Box$

In additive manufacturing systems, materials are extruded layer-wise. Hence, the direction in Z axis should always be in one direction. This allows us to exclude the estimation of direction motor in Z axis.

2.4.2 Leakage Quantification

In the data-driven attack model, reconstruction of the G-code will depend on accurate estimation of the individual parameters of the G-code. Hence, in this chapter, the capability of the data-driven attack model to accurately estimate the individual parameters is focussed. The estimation is done using regression models to predict the continuous speed value, whereas, classification models are used to predict the axis information. Hence, leakage may simply be quantified by measuring the separability of the nozzle movement and prediction accuracy of the nozzle speeds in each axis. The separability is specifically measured using the area under the curve (AUC) of the receiver operating characteristic (ROC) curve, whereas the speed prediction accuracy can be measured using the mean squared error (MSE). The accuracy of the classifiers can be calculated as follows:

$$Accuracy = \frac{TP + TN}{Total\ Sample} \tag{2.10}$$

where $True\ Negative\ (TN)$ and $True\ Positive\ (TP)$ are the total number of right classifications made by the classifier.

2.4.3 Leakage Exploitation

The accuracy with which an adversary is able to exploit the acoustic leakage depends on their ability to estimate the functions $\hat{f}_d(.)$. Breaking the process for estimating $\hat{g}_i$ into multiple estimation functions improves the adversarial attack model by focusing on only those parameters in g_i that are required for breaching the confidentiality of the system. However, the accuracy of the attack model now becomes a function of the successful estimation of individual functions. Moreover, to completely steal the intellectual property inherent in the geometry of a 3D object, each line segment in each layer must be reconstructed with error $e \geq e_P$, where e_P is the error introduced due to process variation. A strong attacker will be able to acquire accuracy with $e = e_P$. However, in a competitive market for products, losing even minute details about the geometry of their product can be disastrous for a company. Hence, in such scenarios, an attacker may just have to infer about the geometry of the 3D objects without 100% accuracy. Based on the domain knowledge of an attacker, they may be able to reconstruct the 3D object with further processing. However, this capability (what kind of domain knowledge an attacker might have) of an attacker is not covered in this chapter.

2.5   Attack Model Description

2.5.1   Attack Model

In the attack model (shown in Fig. 2.4), the intention of an attacker is to steal the geometry details of an object which is one of the intellectual properties for a company using a 3D printer. Even though these details may be stolen by acquiring the 3D object itself, our attack model is most suitable for stealing the IP during the prototyping stage, where 3D printers like the one based on FDM technique are used to visualize the geometry of the company products. An attacker may be a person who has low-level access (can be in a vicinity of the 3D printer) to the 3D printer but not to the digital process chain files (for example, STL, G-codes, etc.) itself. Moreover, they do not have any access to the digital process chain tools and software either. They will have access to a replica model of the target machine that has to have the same physical structure as the target 3D printer. On this replica model, they can perform any sort of prior experiments to train their learning algorithms to estimate the function $\hat{f}_d(.)$. However, the learning should be performed in an environment that is as close to the target machine as possible. The learning algorithms consist of multiple estimated functions $\hat{f}_d(.)$ to predict individual parameters of a G-code. These predicted values are then combined to reconstruct the G-code, and eventually the geometry of an object.

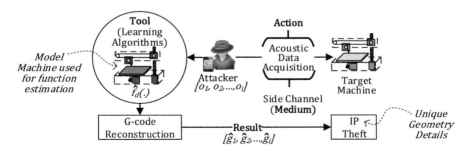

Fig. 2.4 Data-driven acoustic side-channel attack model

2.5.2   Components of the Attack Model

2.5.2.1   Data Acquisition

The components of the attack model is shown in Fig. 2.5. The first step of acquiring the observable analog emissions, $[o_1, o_2, \ldots, o_i]$, is to place an audio recording device such as a mobile phone near the 3D printer. The sampling frequency of the recording device must be higher than 40 kHz to capture the sound in the audible range to avoid aliasing effect [28]. The distance of the audio device from the 3D printer and the angle to the different sources of sound (stepper motor X and stepper motor Y) will also determine the accuracy of leakage exploitation. Moreover, in case of devices that are enclosed, a contact microphone may be utilized to estimate $\hat{f}_d(.)$. This will also reduce the influence of environmental noise on the acquired acoustic emission. For better accuracy, even multiple microphones may be placed to localize and remove environmental noise during the attack.

2.5.2.2   Noise Filtering

A digital finite impulse response bandpass filter is used to eliminate the noise from low-frequency alternating current from the power source, and the high-frequency noise generated by the hybrid stepper motor winding when it is in charged and in idle state. The pass-band frequency for noise removal is between 100 Hz and 20 kHz.

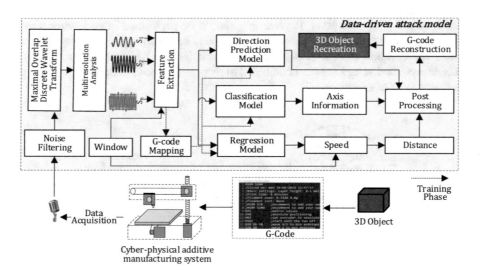

Fig. 2.5 Components of the data-driven attack model

2.5.2.3   Maximal Overlap Discrete Wavelet Transform and Multiresolution Analysis

In [29], a fixed length window was used to extract various time and frequency domain features. With this configuration, feature extraction is more challenging. Capturing smaller movements requires smaller frame sizes to improve the temporal resolution, while improving the speed and frequency resolution requires larger frame sizes. This trade-off prescribes the use of discrete wavelets transforms, to preserve both time and frequency domain features. Maximal overlap discrete wavelet transform (MODWT) is a redundant form of the discrete wavelet transform, where a time series signal is transformed into coefficients related to the variation of a set of scales of the mother wavelet. Due to the added redundancy, MODWT makes it easier for the alignment of the decomposed wavelet and scaling coefficients at each level with the original time series. The MODWT allows us to decompose the time series data into multiple levels of scales and different frequency regions without having to worry about the length of the window. These signals are then passed through the multiresolution analysis block to decompose it into multiple signals $(S_1, S_2, \ldots, S_n)$ for feature extraction. In [29], uniform frequency scales are used to extract features using *short-term Fourier transform* (STFT). Even though, *mel-frequency cepstral coefficients* (MFFC) are used in [29], which uses non-uniform scaling focusing on higher number of frequency features in the lower frequency range, it is found that this feature extraction is not feasible for the 3D printer acoustic analysis due to variation of frequency distribution based on the travel feed-rate. Hence, analysis of signals decomposed by MODWT is performed to define the non-uniform frequency scales for feature extraction. Moreover, five levels of MODWT signals are computed to define the non-uniform frequency scales.

2.5.2.4   Feature Extraction

Features commonly used in speech pattern recognition [30] in the time and frequency domains are used to train the learning algorithms. In the time domain, the features extracted are *frame energy*, *zero crossing rate (ZCR)*, and *energy entropy* [30]. The features extracted from the frequency domain are *spectral entropy* and *spectral flux*. Multiple signals $(S_1, S_2, \ldots, S_n)$ are obtained from the multiresolution analysis block. For each signal, varying number of frequency scales are allocated to extract spectral energy values. Two frame sizes, 50 ms and 20 ms, are used for testing the performance of the attack models with MODWT and STFT based features. From each frame, features are extracted to create a feature vector and supply it to the training algorithm. For a given frame of length F_L with audio signals $x(i) = 1, 2, \ldots, F_L$, different features are extracted as follows:

$$Frame\ Energy\ (E) = \sum_{i=1}^{F_L} \mid x(i) \mid^2 \qquad (2.11)$$

Frame energy is enough to predict direction when the printer is only printing in one axis; however, spectral energy is required while predicting the direction in multiple axes movement. ZCR is calculated as follows:

$$ZCR = \frac{1}{2F_L} \sum_{i=1}^{F_L} \mid sign[x(i)] - sign[x(i-1)] \mid^2 \qquad (2.12)$$

ZCR is high when the printer is not making any sound, due to the noise, and low when it is printing. For energy entropy, the frame is divided into short frames of length K. If E_j is the energy of the jth short frame, then:

$$Energy\ Entropy = -\sum_{j=1}^{K} e_j log_2(e_j), \quad e_j = \frac{E_j}{\sum_{i=1}^{K} E_i} \qquad (2.13)$$

Energy entropy measures the abrupt change in the energy of the signal, and may be used to detect the change of motion. For frequency domain data, let $X_i(k)$, $k = 1, \ldots, F_L$ be the magnitude of the fast Fourier transform (FFT) coefficient of the given frame. For spectral entropy, the spectrum is divided into L sub-bands. Let E_f be the energy of the fth sub-band then:

$$Spectral\ Entropy = -\sum_{f=1}^{L-1} n_f log_2(n_f), \quad n_f = \frac{E_f}{\sum_{j=1}^{L-1} E_j} \qquad (2.14)$$

Spectral flux measures spectral change between two successive frames, and may be used to detect the change of speed of the nozzle while printing within each layer.

$$Spectral\ Flux_{i,i-1} = \sum_{k=1}^{F_L}(EN_i(k) - EN_{i-1}(k))^2, \quad EN_i(k) = \frac{X_i(k)}{\sum_{j=1}^{F_L-1} X_i(j)}$$
$$(2.15)$$

2.5.2.5   Regression Model

The regression model consists of a collection of models, each using a supervised learning algorithm for regression as shown in Fig. 2.6. These models are used for estimating the functions $v_{x_i} = \hat{f}(o_i, \alpha)$ and $v_{y_i} = \hat{f}(o_i, \alpha)$. These functions are used to extract information about speed in X direction given only one axis movement, and speed in X direction given the motion in two axis. Similarly, this is done for speed in Y direction as well.

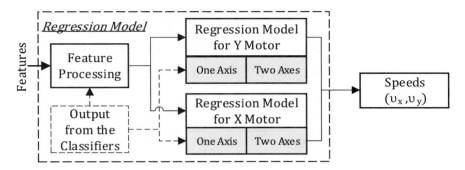

Fig. 2.6 Regression model for motor speed prediction

Assumption The speed in the Z direction while printing the given model with the given printer is fixed and the speed of extrusion can be calculated as a function of layer height and nozzle diameter. □

For a given 3D printer, the layer height is assumed to be fixed. This relaxes the complexity for leakage exploitation by reducing the need for estimating speeds v_x and v_e. The speed of the printing, also known as the *travel feedrate*, is determined by training these regression algorithms [31]. After gaining the information about the *travel feedrate*, the distance moved by the nozzle can be calculated as follows:

$$Distance = Framesize \text{ (ms)} \times Speed \text{ (mm/ms)} \qquad (2.16)$$

When the nozzle is moving in only one axis, the regression model may just take the features directly without further processing; however, when the nozzle is moving in two or more axes, the audio signal from one motor is combined with the others. Hence, it becomes imperative to separate these signals before the regression model can be used to predict the speed. Algorithm 1 provides the pseudo code for performing the spectral subtraction necessary when motion is involved in both the X and Y axes. It takes features, extracted from the audio when both the X and Y motors are running, and the features from the training phase for individual motor X and Y as the input. Spectral subtraction is not performed for Z motor because it only moves one layer at a time and the distance it moves is normally fixed for a given object. While training, n number of speeds, in incremental number is taken to train the regression models. For each of these speeds, lines 1 and 2 calculate the average magnitude of spectral features. Then, for each of the speeds, line 6 assumes the speed of the Y motor and the spectral components are subtracted from the combined spectral features of X and Y. By subtraction, the spectral components present in Y is removed from the combination of these features. Line 7 gives the predicted speed for the given value of speed in the Y direction. This speed is used to subtract the spectral features of X in the particular speed and again use this value to predict the speed for motion in Y axis. In lines 11 and 12, the speed of X and Y that gives the minimum difference in the predicted speed and output speed in Y axis is chosen as an output.

Algorithm 1: Feature processing and speed calculation with motion in XY axes

---

Input: Feature Vectors $xy_\beta, x_\beta, y_\beta$ $// \; \beta \to$ Total features
Output: Speed ϑ_x, ϑ_y $// \; \vartheta \to$ Speed

1 $\vartheta^i_{x_{mean_{\beta j}}} = \frac{1}{N_i} \sum_{n=1}^{N_i} x_{\beta_{n.j}} \; // \; j = 1 \longrightarrow \beta n \;\; N_i \longrightarrow Total \; data \; for \; speed \; i$

2 $\vartheta^i_{y_{mean_{\beta j}}} = \frac{1}{N_i} \sum_{n=1}^{N_i} y_{\beta_{n.j}}$

3 **for** $each \; xy$ **do**

4   **for** $\vartheta^i \; in \; range(v^1, v^n)$ $// \; n:$ Total number of speed used in training

5   **do**

6          $xy_{beta(xy-y)} = xy_{beta} - \vartheta^i_{y_{mean_\beta}}$

7          $\vartheta^i_x \leftarrow RegressionModel1(xy_{beta(xy-y)})$

8          $xy_{beta(xy-x)} = xy_{beta} - \vartheta^i_{x_{mean_\beta}}$

9          $\vartheta^i_y \leftarrow RegressionModel2(xy_{beta(xy-x)})$

10         $diff_i =| \vartheta^i_y - \vartheta^i |$

11     $\vartheta_y = \vartheta^i_y with \; minimum \; diff_i$

12     $\vartheta_x = \vartheta^i_x with \; minimum \; diff_i$

13 **return** ϑ_x, ϑ_y

2.5.2.6 Classification Model

As shown in Fig. 2.7, to determine the axis in which the nozzle is moving, the classification model consists of a collection of classifiers to convert the classification problem into two-class separation model. This in fact will estimate the functions $a_{x_i} = \hat{f}(o_i, \alpha), a_{y_i} = \hat{f}(o_i, \alpha)$, and $a_{z_i} = \hat{f}(o_i, \alpha)$. It is found that this model gives better prediction results than multi-class classifier models. Each of these classifiers consists of supervised learning algorithms for classification. Algorithm 2 gives the pseudo code which takes the output from the classifiers to determine the axis of movement. It also gives information such as whether the layer has changed or not, and whether the nozzle is moving in X and Y axis with the same or different speed (Fig. 2.8).

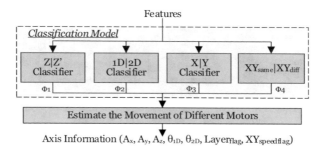

Fig. 2.7 Classification model for axis prediction

Algorithm 2: Estimate the axis of movement

Input: Classifier Outputs ϕ_1, ϕ_2, ϕ_3, ϕ_4
Output: Axis Parameters A_x, A_y, A_z, Θ_{1D}, Θ_{2D}, $Layer_{flag}$, $XY_{speedflag}$

1 $\Theta_{1D} = 0$, $\Theta_{2D} = 0$ // $A \longrightarrow axis$, $\Theta \longrightarrow dimension$
2 $Layer_{flag} = 0$, $XY_{speedflag} = 0$
3 $A_x = 0$, $A_y = 0$, $A_z = 0$ // Initialize to zero
4 **if** $\phi_1 == 1$ **then**
5      $Layer_{flag} = 1$, $A_z = 1$

6 **else**
7   **if** $\phi_2 == 1$ // One dimension movement
8   **then**
9         $\Theta_{1D} = 1$
10   **if** $\phi_3 == 1$ // Movement in X axis
11   **then**
12          $A_x = 1$

13   **else**
14          $A_y = 1$ // Movement in Y axis

15   **else**
16        $\Theta_{2D} = 1$, $A_x = 1$, $A_y = 1$
17   **if** $\phi_4 == 1$ // X and Y move with same speed
18   **then**
19          $XY_{speedflag} = 1$

20   **else**
21          $XY_{speedflag} = 0$ // Different speed

22 **return** A_x, A_y, A_z, Θ_{1D}, Θ_{2D}, $Layer_{flag}$, $XY_{speedflag}$

2.5.2.7 Direction Prediction Model

Most of the 3D printers have motors in a fixed location. However, the base plate, the nozzle or combination of both are always in motion while printing. Therefore, vibration is conducted from the motor to the nozzle and the base plate of the printer. This means that the audio source physically gets closer or away from the recording

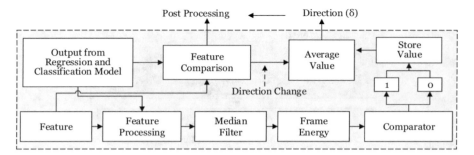

Fig. 2.8 Direction prediction model

device while printing. The frame energy of the audio signal can be used to check the direction of motion. For multiple motor movements, the difference of feature in the frequency domain is used to calculate the energy of only those spectral components that represent the specific motor. In order to suppress the high fluctuation, median filtering is applied to the sequence of frame energies to smooth the curve of frame energies. The prediction model will output 1 if the frame energy is increasing and 0 if the frame energy is decreasing. In order to aid the direction prediction model and the post-processing, a feature comparison block measures the distance (euclidean distance) between consecutive frame features. If the motion of direction changes, then there is a large difference in the features between the consecutive frames. This spike is used to detect the change in the direction of motion of the nozzle.

2.5.2.8   Model Reconstruction

For reconstructing the G-code, it is necessary to determine whether the 3D printer nozzle is actually extruding the filament or not. From the analysis, it was found that the printer nozzle moves at a higher speed when it is not extruding the filament. Hence, determining whether it is printing or not printing becomes a task of finding out the speed at which the nozzle is moving. This information is acquired from the regression model. The extrusion amount for a given segment is machine-specific, and can be calculated as a function of the layer height, and the nozzle diameter. After acquiring the output from the regression model, classification model, and direction prediction model, Algorithm 3 calculates the positive or negative distance movement. Finally, Algorithm 4 reconstructs the G-code for the printed object.

2.5.2.9   Post-Processing for Model Reconstruction

There is high mutual information between the G-code and the sound retrieved from the physical medium. For G-codes, let G be a discrete random variable with $f(g)$ as its probability distribution function at g. Let O be a discrete random variable representing the feature extracted from the acoustics with $f(o)$ as its probability

---

Algorithm 3: Calculate distance moved in each axis, and check extrusion

---

Input: Output from Classifier and Regression Models $\vartheta_x, \vartheta_y, \vartheta_z, \mathbf{w}, A_x, A_y, A_z, \delta_x, \delta_y, \delta_z$
Output: Distance Values d_x, d_y, d_z, d_E // $d \rightarrow Distance, x_\vartheta \rightarrow Speed\ in\ X\text{-}axis$
1 $d_x = 0, d_y = 0, d_z = 0$ // $w \rightarrow Frame\ length, A \rightarrow AxisFlag, \delta \rightarrow Direction$
2 **for** *each i in x, y, z* **do**
3   **if** $A_i == 1$ // Axis flag set
4   **then**
5   **if** $\delta_i == 1$ **then**
6              $d_i = \vartheta_i \times w$ // Positive distance
7   **else**
8              $d_i = \neg\vartheta_i \times w$ // Negative distance

9 **if** $\vartheta_x >= Speed_{High} \parallel \vartheta_y >= Speed_{High}$ **then**
10      $d_E = 0$ // No extrusion in high speed
11 **else**
12      $d_E = e_d$ // $e_d \longrightarrow Machine\ specific\ extrusion$
13 **return** d_x, d_y, d_z, d_E

Algorithm 4: Generate G-code of the object

Input: Distance and Frame Length d_x, d_y, d_z, d_E , w
Output: *G-code*   // Initialize to zero
1 $dr_x = 0, dr_y = 0, dr_z = 0, dr_E = 0$

2 $\vartheta = \frac{\sqrt{d_x^2 + d_y^2 + d_z^2}}{w}$ // Travel feedrate
3 $dr_x = dr_x + d_x$ // Distance moved in X axis
4 $dr_y = dr_y + d_y$ // Distance moved in Y axis
5 $dr_z = dr_z + d_z$ // Distance moved in Z axis
6 $dr_E = dr_E + d_E$ // Extrusion amount
7 *G-code* $\leftarrow$ G1 $F(\vartheta)\ X(dr_x)\ Y(dr_y)\ Z(dr_z)\ E(dr_E)$
8 **return** *G-code*

---

distribution function. Then the entropy of each of these random variables may be given as:

$$H(G) = -\sum_{g\in G} f(g)log_2 f(g) H(O) = -\sum_{o\in O} f(o)log_2 f(o) \qquad (2.17)$$

If $f(g, o)$ and $f(g|o)$ are the joint and conditional probabilities of the random variables, respectively, then the conditional entropy $H(G|O)$ is calculated as:

$$H(G|O) = -\sum_{o\in O}\sum_{g\in G} f(g, o)log_2 f(g|o) \qquad (2.18)$$

The conditional entropy measures the amount of information required to describe the outcome of a random variable G, given the information about a random variable O. In this context, in addition to the information gathered from O, the amount of additional additive manufacturing context-based information required to reconstruct the G-code is directly related to the mutual information. This is calculated as:

$$I(G; O) = H(G) - H(G|O) \tag{2.19}$$

It is found that the uncertainty of reconstruction of G-code or the entropy $H(G|O)$ increases when the distance of the microphone is further away from the printer or when there is added noise in the environment. It also increases when the speed of the printer is high and there are more short and rapid movements. During these scenarios, we can use the properties of additive manufacturing to post-process the data achieved from the learning algorithms. Specifically, we have used two post-processing stages which utilize specific additive manufacturing context-based information.

Post-Processing Stage I In this stage, $H(G|O)$ is reduced by utilizing the fact that until the change of motion occurs, the nozzle moving in one particular dimension with a particular speed has a similar feature vector. By taking the output from the feature comparison model, the acquired acoustic data is segmented into sections with a similar movement. In this post-processing stage, then the output of the classifiers is chosen to be the highest occurring value in the given segment, and for regression, the speed obtained is averaged within the same section. This is similar to averaging used in digital signal processing to increase the signal-to-noise ratio (SNR).

$$SNR_{dB} = 10 log_{10} \left(\frac{Power_{signal}}{Power_{Noise}} \right) \tag{2.20}$$

When the SNR is increased, the entropy of the signal is reduced. As there is a high correlation among the features extracted from successive frames of the audio collected from the 3D printer, averaging the output of the classification and the regression model increases the SNR and thus reduces $H(G|O)$.

Post-Processing Stage II After applying post-processing stage I, the second stage measures the similarity between the two layers. The similarity of two layers is measured in terms of the number of segments, the sequence of motions in each layer, and the length of each segment. This post-processing stage helps the attack model in reducing the error due to the miscalculated direction and fluctuating lengths by taking the average of segment lengths and direction among the similar layers of the 3D object.

2.5.3   Attack Model Training and Evaluation

Our testbed, shown in Fig. 2.9, consists of a Printrbot 3D printer [32] with open source *marlin* firmware. It has four stepper motors. Motion in the X axis is achieved by moving the base plate, whereas the nozzle itself can be moved in the Y and Z directions. The audio is recorded using a cardioid condenser microphone (Zoom H6) [33], which has a sampling frequency of 96 kHz and stores the data at 24 bit per sample. The audio recorder is placed within 20 cm of the 3D printer. From the experiments, it is observed that for the direction prediction model to work efficiently, the audio device has to be placed at 45° angle to both the X and the Y axes as shown in Fig. 2.9. This allows the audio device to capture the variation of sound in both X and Y directions. The digital signal processing, feature extraction, and post-processing are performed in MATLAB [34], whereas the training of learning algorithms, their evaluation and testing are done using Python [35]. The attack model consists of supervised learning algorithms described in Sect. 2.5.2. For training these algorithms, initial training data has to be determined. The training data consists of G-code to move the printer nozzle at different speeds (500–4500 mm/min) and different axes. The speed range chosen is specific to the 3D printer. The G-code for training phase consists of movement in just one axis (X, Y, and Z), two axes (XY, XZ, and YZ), and all three axes (XYZ). The audio signal corresponding to each of these G-codes is pre-processed and labeled for training the learning algorithms. The total length of audio recorded for training is 1 h 48 m. The total number of features extracted is 328 for the window size of 50 ms and 318 for the window size of 20 ms. For regression model, *decision trees* boosted using *gradient boosting* algorithm is used, whereas for the classification model *decision tree classifier* boosted using *AdaBoost* algorithm [31] is used. The learning algorithms are trained by performing K-fold cross validation, with $k = 4$, to test the efficiency of the learners as well as to avoid over or under fitting of the learning algorithms. In the experiments, the regression model is trained only for the nozzle movements in the X and Y directions.

Fig. 2.9 Setup for training and testing

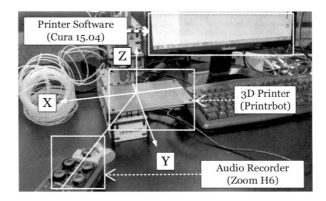

Classification Models Table 2.1 shows the accuracy of the various classifiers. The MODWT and STFT based features extraction are compared with the features extracted in our previous work [29]. It can be observed that compared to the accuracy of the previous classifiers, the accuracy of the MODWT and STFT features based classifiers is higher. Even when the window size is reduced to 20 ms, the accuracy of the classifiers is still higher than the classifiers using features described in [29]. For measuring the accuracy of the classifiers, receiver operating characteristics (ROC) curves are also analyzed. The classifiers capability to separate the two classes is high if the graph lies closer to the upper right corner. This region corresponds to 100% sensitivity (zero false negatives) and 100% specificity (zero false positives). As the collection of classifiers are arranged in a hierarchy, the bottleneck in terms of accuracy is the longest path followed while making the decision. In this case, the longest path involves all the classifiers. From Fig. 2.10a–d, it can be observed that the classifiers have a high sensitivity and specificity with a high area under the curve (AUC). It means the different classes can be accurately classified based on the observed leakage from the side-channel. The AUC for the

Table 2.1 Accuracy of the classification models

Classification model	Classifying	Accuracy (%) [29] win 50 ms	Accuracy (%) (MODWT and STFT) win 50 ms	Accuracy (%) (MODWT and STFT) win 20 ms
ϕ_1	Z or $\sim Z$ axis	99.86	99.9559	99.9400
ϕ_2	$1D$ or $2D$ axis	99.88	99.9700	99.9606
ϕ_3	X or Y axis	99.93	99.9910	99.9778
ϕ_4	XY_{same} or $XY_{\text{different}}$	98.89	99.4644	99.2349

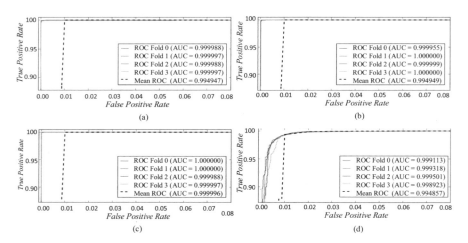

Fig. 2.10 Receiver operating characteristics curve for classifiers. (**a**) ROC for Z or $\sim Z$ (with 20 ms window). (**b**) ROC for $1D$ or $2D$ (with 20 ms window). (**c**) ROC for X or Y (with 20 ms window). (**d**) ROC for XY_{same} or $XY_{\text{different}}$ 2D (with 20 ms window)

classifier classifying whether the movement X and Y axes have the same speed or different speed is comparatively less than other AUCs. This is intuitive as, in multiple axis movement, separation of individual movement is difficult.

Regression Models The accuracy of the regression model is measured in terms of mean squared error (MSE) with the data normalized with zero mean and unit variance. The mean absolute error is presented to understand how the speed prediction varies from the real speed. From Table 2.2, it can be seen that the MSE with new features is less than compared to the MSE of the regression models used in our previous work [29]. With the new features, lower MSE is achieved when the window size is reduced to 20 ms. Lower window size allows us to have higher 3D object reconstruction accuracy when the dimension of the 3D object being printed is lower. It can be seen that the MSE is relatively higher for the value predicted by the regression model for the motion in Y axis when the motion is occurring at two axes. However, this error can be removed during the post-processing stage as the travel feed-rate is generally similar between consecutive frames in each layer of printing. Figure 2.11 shows that there is a linear relationship between the real speed and the predicted speed computed by the regression model. Figure 2.12 shows the feature comparison conducted for the audio recorded while the 3D printer is printing an

Table 2.2 Accuracy of the regression models

Regression model	Movement axis	MSE [29] normalized (win 50 ms)	MSE normalized (MODWT and STFT) (win 50 ms)	Mean absolute error (win 50 ms) (MODWT and STFT) mm/min	Mean absolute error (win 20 ms) (MODWT and STFT) mm/min
X	Only X	0.00616	0.00292	5.876	8.6786
Y	Only Y	0.01874	0.01201	18.3721	23.5821
X	X and Y	0.1658	0.04641	91.4723	140.045
Y	X and Y	0.4290	0.25120	268.07	294.0423

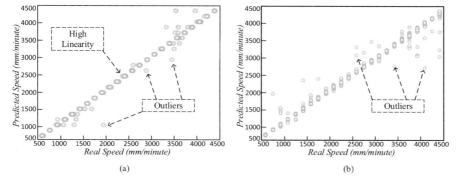

Fig. 2.11 Prediction results for regression models. (**a**) Speed Prediction for X with one axis motion. (**b**) Speed prediction for Y with one axis motion

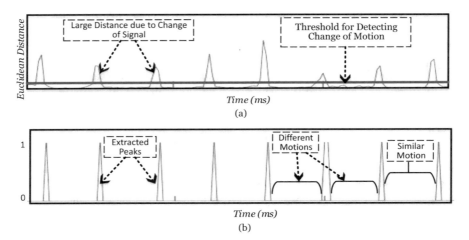

Fig. 2.12 Features segmented with varying motion. (**a**) Distance between features of consecutive frames. (**b**) Peak extracted from the distance measurement

object. It can be observed that when the nozzle changes its direction by analyzing the distance of features between successive frames. Peaks are extracted by applying the threshold obtained during the training phase. A value higher than the threshold is 1, otherwise 0.

2.6 Results for Test Objects

In order to test our attack model, various benchmark parameters which affect the accuracy of the attack model are defined. While printing, multiple similar 2D layers are printed to achieve a 3D object. Hence, our test objects also consist of 3D objects with simple 2D geometry (such as square, triangle, etc.) repeated over multiple layers. The various benchmark parameters in designing these test objects are as follows:

2.6.1 Speed of Printing

The fixed frame rate affects the temporal and spectral features extracted from the audio. With the increase in the speed, faster rate of change of spectral features will not be captured and this can degrade the performance of the attack model. Hence, the speed of printing is varied to test the accuracy of the attack model.

2.6.2   The Dimension of the Object

With smaller objects, shorter nozzle movements are present. To represent these shorter movements, the temporal resolution of the features is increased by making the frame size smaller. To test our attack model with smaller objects, we vary the size of the object being printed.

2.6.3   The Complexity of the Object

Printing a complex object incorporates movement in more than one axis. Hence, to increase the complexity of the object being created, we have tested the acoustic model with shapes consisting of simultaneous multiple axis movements, such as a triangle.

For each of the test objects, to test the reconstruction capability of the attack model for 3D objects, we have printed it in five layers. Since the attack model uses post-processing to remove any layer that is not overlapping with the previous layers, having larger number of layers with same base 2D outline throughout would increase the accuracy of the attack model. Hence, we have chosen smaller number of layers as a proof of concept, and for demonstration purpose. In order to provide the result in a meaningful manner, instead of calculating the mean squared error, the mean absolute percentage error (MAPE) is used for the distance prediction.

$$MAPE = \frac{1}{n} \sum_{t=1}^{n} \left| \frac{A_t - P_t}{A_t} \right| \tag{2.21}$$

where A_t is the actual speed and P_t is the speed predicted by the attack model. Since the frame size (20 ms) is same for all the features, the distance calculation error is also be given by Eq. 2.21. Table 2.3 provides results for the different parameters used to test the accuracy of the attack model. The average classification accuracy and regression MAPE before the post-processing stage are 74.43% and 13.17%, respectively. This is an improvement over the results of our previous work [29], where the classification accuracy is 66.29% and regression MAPE is 20.91%. After the post-processing stages, the classification accuracy is 86%, and the regression MAPE is 11.11%. This result is also an improvement over the classification accuracy of 78.35% and regression MAPE of 17.82% after post-processing stages presented in our previous work [29] due to better feature extraction achieved by combination of MODWT and STFT.

Table 2.3 Test results for square and triangle

Dimension (mm)	Speed (mm/min)	Regression MAPE (%)	Classification accuracy (%)	Classification accuracy App I (%)	Classification accuracy App II (%)	Regression MAPE App II (%)
Square (side)						
20	900	4.83	92.56	99.51	99.51	3.39
	1200	6.80	91.82	97.68	97.68	5.14
	1500	9.37	85.91	89.34	92.12	7.65
	1700	15.47	82.43	88.27	88.27	11.21
10	900	4.97	89.28	98.55	98.55	3.67
	1200	7.58	83.66	87.81	88.61	6.79
	1500	11.88	78.81	86.11	90.23	9.81
	1700	18.67	74.93	82.12	85.31	14.32
5	900	8.72	76.72	84.63	86.62	5.61
	1200	12.64	69.44	76.22	76.22	8.91
	1500	17.44	62.05	74.07	74.07	12.57
	1700	23.91	57.29	62.64	62.64	18.99
Triangle (base, height)						
30, 20	900	4.48	90.85	98.71	98.71	4.46
	1200	5.19	88.70	95.06	96.34	5.11
	1500	6.85	85.15	91.18	91.18	6.73
	1700	10.33	78.77	84.30	84.30	9.82
20, 20	900	5.45	87.50	94.77	97.35	4.80
	1200	7.41	84.01	90.14	90.14	6.37
	1500	9.22	79.99	85.61	85.61	8.88
	1700	16.02	77.41	79.39	83.13	14.25
10, 5	900	16.57	70.84	78.42	82.37	14.77
	1200	21.43	68.23	74.90	74.90	19.69
	1500	32.78	63.93	72.28	74.44	28.55
	1700	38.41	58.51	65.71	65.71	35.21
Average		**13.17**	**74.43**	**84.89**	**86.00**	**11.11**
Average [29]		**20.91**	**66.29**	**76.04**	**78.35**	**17.82**

The bold letters presents the average result. It is highlighted to present the summary of the table.

2.6.4   Reconstruction of a Square

A square incorporates movements of stepper motors in all axes, however, one at a time. From Table 2.3, it can be seen that the accuracy of the classifier for reconstructing the G-code is as high as 92.56% with MAPE of just 4.83%. After post-processing stages the same accuracy has been increased to 99.51% for the classifier with MAPE of just 3.39%. It can also be observed that as the *travel feedrate* increases to 1700 mm/min, the accuracy of the classifier and the regression model decreases. Also for short movements such as 5 mm, the accuracy of the attack

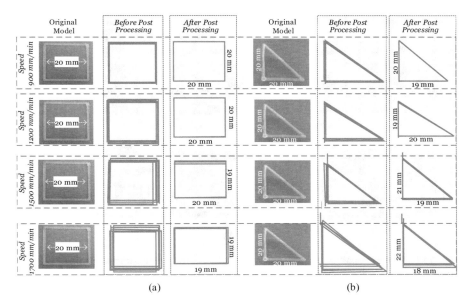

Fig. 2.13 Reconstructed test objects. (**a**) Square reconstruction. (**b**) Triangle reconstruction

model decreases. Figure 2.13a shows the square reconstructed by the attack model for a side length of 20 mm.

2.6.5 Reconstruction of a Triangle

While constructing a triangle, both the X and Y stepper motors move, and this affects the reconstruction accuracy of the attack model. From Table 2.3, it can be seen that the accuracy of the classifier of the attack model is as high as 90.85% with MAPE of just 4.48% before post-processing and after post-processing, they are 98.71% and 4.46%, respectively. As expected, the accuracy of the learning algorithms decreases with increasing speed and decreasing length of the movement. Also, classification and regression accuracy of reconstructing the triangle is less when compared to the square. Figure 2.13b shows the reconstructed shape of the triangle. For higher travel feed-rates, the accuracy of the reconstruction is lower.

2.6.6 Case Study: Outline of a Key

As a case study, to test the attack model against a combination of shapes, an object representing the outline of a key at 900 mm/min *travel feedrate* is printed. The classification accuracy obtained for this object before post-processing is 88.83%

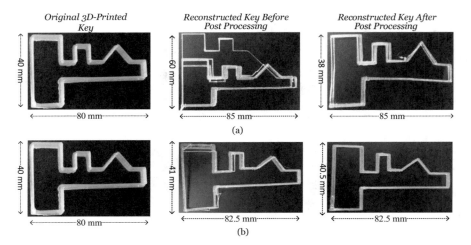

Fig. 2.14 Reconstruction of a key as a case study. (**a**) Key reconstructed in [29]. (**b**) Key Reconstructed in with new features

and the regression MAPE is 5.62%. This is better than the results presented in our earlier work [29], where the classification accuracy obtained before the post-processing is 83.21% and the regression MAPE is 9.15%. After the post-processing, the classification accuracy obtained is 96.32% and the regression MAPE obtained is 3.92%. This is also an improvement over the result presented in our previous work [29], where the classification accuracy obtained is 92.54% and the regression MAPE obtained is 6.35%. The object reconstructed by the attack model is shown in Fig. 2.14. As can be seen, before post-processing stage II, there are some non-uniform lengths in each of the layers of the object. However, after post-processing stage II, these errors are corrected. In terms of dimension, it can be observed that the reconstructed key varies in length and width compared to the original object. Nevertheless, the general outline of the key is reconstructed accurately. Moreover, the accuracy in terms of the length obtained after the post-processing stage is 92.48%, which is calculated by dividing the difference between the original length and the predicted length of each segment of each of the layer by the total length of all the segments in all the layers. This is better than the results presented in [29], where the accuracy is 89.72%. On increasing the travel feed-rate for the given test case, it can be expected to have higher distortion as demonstrated by the results for the reconstruction of test objects such as square and triangle, where increasing the speed caused drastic misalignment of layers, error in direction prediction, and imprecise dimension prediction.

2.7 Discussion

2.7.1 Technology Variation

In this chapter, the experiment was performed using an FDM technology based 3D printer. The key assumption for the attack model is that there is a correlation between the G-code and the radiated sound, and an attacker is capable of acquiring these radiated signals. With this, any 3D printer that is capable of suppressing the acoustic emission from the printer can effectively avoid this kind of attack. For example, other 3D printing technologies such as stereolithography (SLA), selective laser melting (SLM), electron beam melting (EBM), etc., that uses UV-laser, high beam laser, electron beam etc., to harden either the liquid resin or metal powder, have minimum number of components that can generate acoustic emissions. In such scenarios, different analog emissions (magnetic, power, thermal, etc.) may be used for breaching the confidentiality of the system.

2.7.2 Sensor Position

In the experiment, the sensor is placed at a fixed position (20 cm) from the 3D printer. However, the sensor position plays an important role in the accuracy of the system. For instance, if the sensor is placed on the top parallel to Z axis of the 3D printer, the variation in X and Y axes direction will be difficult to capture. A better sensor position exploration can be done to understand how the position affects the accuracy of the attack model.

2.7.3 Sensor Number

In this chapter, only a single acoustic sensor is used to acquire the acoustic emissions. However, a motivated attacker might have multiple sensors placed around the 3D printer to acquire the signals. This may change the accuracy of the attack model. It will help in better localization of the source of acoustic, and remove other environmental noise (for example, using blind source separation method). However, the angle and distance between these sensors have to be explored before it can be used for data acquisition.

2.7.4 Dynamic Window

Due to the fixed frame size incorporated for feature extraction in our experiments, the accuracy of the attack model is reduced for higher speeds and smaller dimensions. In order to capture smaller movements, the temporal resolution of the features extracted has to be increased by making the frame size smaller. However, for faster speeds, larger frame size is necessary to increase the frequency resolution for better spectral features. This trade-off dictates that an adaptive frame size needs to be incorporated to increase the accuracy of the attack model.

2.7.5 Feature Separation during Multiple Axis Movement and Noise

The separation of the sound source from a combination of sound is a well-known problem in speech processing. In our attack model, spectral subtraction is incorporated to acquire features that are unique to each of the stepper motors. However, there are other sound separation methods [36, 37]. Moreover, only one audio sensor is used for separating two features. Incorporating two or more audio sensors may improve the results further.

2.7.6 Target Machine Degradation

Over a longer period of time, due to mechanical degradation, the vibration produced by the 3D printer will vary compared to the new models. Since an attacker cannot access the target model for estimating the leakage function, they might have less accuracy in stealing the information. However, one possible solution to tackle this issue would be to continuously update the model function using a 3D printer model that closely handles the same workload as is done in the industry through accelerated aging test methods, and capture the degradation trend to isolate any noise caused by the mechanical wear and tear that does not aid in information leakage.

2.8 Summary

This chapter introduced and demonstrated a novel data-driven acoustic side-channel attack model on cyber-physical additive manufacturing systems. An analysis on sources of acoustic emission in a fused deposition modeling technique based 3D printer is presented, and leakage analysis is performed to highlight the parameters of G-code that can be inferred from the acoustic side-channel. Maximum overlap discrete wavelet transform is incorporated in feature extraction to acquire better G-code and 3D object reconstruction results than short-term Fourier transform and Mel-frequency cepstral coefficient features based attack model. The validation of our novel attack model with a state-of-the-art 3D printer shows that objects with different benchmark parameters such as speed, dimension, and complexity can be reconstructed using acoustic side-channel attacks. Our experiments show an average axis prediction accuracy of 74.43%, and average length prediction error of 13.17%. Furthermore, with post-processing, a high average axis prediction accuracy of 86% and average length prediction error of 11.11% is achieved. As it is explained in the discussion section, there are several research challenges that remain open. The work presented in this chapter serves as a proof of concept for the possibility of physical-to-cyber attacks on cyber-physical additive manufacturing systems.

References

1. Leukers, B., Gülkan, H., Irsen, S. H., Milz, S., Tille, C., Schieker, M., et al. (2005). Hydroxyapatite scaffolds for bone tissue engineering made by 3D printing. *Journal of Materials Science: Materials in Medicine, 16*(12), 1121–1124.
2. ISS Platform and Feedstock Recycling. (2014). NASA advanced manufacturing technology.
3. Short, B. (2015). Quality metal additive manufacturing (QUALITY MADE) enabling capability. www.navy.mil.
4. NDIA. (2014). Cyber security for advanced manufacturing. Technical report, National Defense Industrial Association.
5. Reznick, C. (2015). Manufacturing: A persistent and prime cyber attack target. www.cohnreznick.com.
6. Yampolskiy, M., Andel, T. R., McDonald, J. T., Glisson, W. B., & Yasinsac, A. (2014). Intellectual property protection in additive layer manufacturing: Requirements for secure outsourcing. In *Proceedings of the 4th Program Protection and Reverse Engineering Workshop* (New York, NY: ACM).
7. Gibson, I., Rosen, D. W., & Stucker, B. (2014). *Additive manufacturing technologies*. Berlin: Springer.
8. Ashford, W. (2014). 21 percent of manufacturers hit by intellectual property theft.
9. Wall, D. S., & Yar, M. (2010). Intellectual property crime and the internet: cyber-piracy and stealing information intangibles. In *Handbook of internet crime* (p. 255).
10. Branigan, T. (2010). Google to end censorship in china over cyber attacks. In *The Guardian* (p. 01-12).
11. Standaert, F.-X., Malkin, T. G., & Yung, M. (2009). A unified framework for the analysis of side-channel key recovery attacks. In *Annual International Conference on the Theory and Applications of Cryptographic Techniques*. Berlin: Springer.
12. Chhetri, S. R., Canedo, A., & Faruque, M. A. A. (2018). Confidentiality breach through acoustic side-channel in cyber-physical additive manufacturing systems. *ACM Transaction on Cyber-Physical Systems, 2*(1), 3.
13. SketchUp Make. (2015). www.sketchup.com.
14. Adobe Photoshop CC. (2015). www.adobe.com.
15. Holbrook, T. R., & Osborn, L. S. (2014). Digital patent infringement in an era of 3D printing. *UC Davis Law Review, 48*, 1319.
16. Sturm, L. D., Williams, C., Camelio, J., White, J., & Parker, R. (2014). Cyber-physical vulnerabilities in additive manufacturing systems. *Context, 7*, 8.

17. Backes, M., Dürmuth, M., Gerling, S., Pinkal, M., & Sporleder, C. (2010). Acoustic side-channel attacks on printers. In *USENIX Security Symposium* (pp. 307–322).
18. Toreini, E., Randell, B., & Hao, F. (2015). *An acoustic side channel attack on enigma.*
19. Davis, A., Rubinstein, M., Wadhwa, N., Mysore, G. J., Durand, F., & Freeman, W. T. (2014). The visual microphone: Passive recovery of sound from video. *ACM Trans. Graph, 33*(4), 79
20. Vincent, H., Wells, L., Tarazaga, P., & Camelio, J. (2015). Trojan detection and side-channel analyses for cyber-security in cyber-physical manufacturing systems. *Procedia Manufacturing, 1*, 77–85.
21. Hughes, A., & Lawrenson, P. J. (1975). Electromagnetic damping in stepping motors. In *Proceedings of the Institution of Electrical Engineers* (Vol. 122, pp. 819–824). London, UK: IET.
22. Kenjō, T., & Sugawara, A. (1994). *Stepping motors and their microprocessor controls.* Oxford: Oxford University Press.
23. Yang, S. J. (1981). *Low-noise electrical motors* (Vol. 13). Oxford: Oxford University Press.
24. Heller, B., & Hamata, V. (1977). *Harmonic field effects in induction machines.* Amsterdam: Elsevier.
25. Gieras, J.F., et al. (2005). *Noise of polyphase electric motors.* Boca Raton, FL: CRC Press.
26. Timár-P, L. T.-P., & Tímár, P. L. (1989). *Noise and vibration of electrical machines* (Vol. 34). Amsterdam: North Holland.
27. So, E. C. T., Williams, R. G. D., & Yang, S. J. (1993). ECT So, RGD Williams, and SJ Yang. A simple model to calculate the stator radial vibration of a hybrid stepping motor. In *Conference Record of the 1993 IEEE Industry Applications Society Annual Meeting* (pp. 122–129). Piscataway, NJ: IEEE.
28. Schafer, R. W., & Oppenheim, A. V. (1989). *Discrete-time signal processing* (Vol. 2). Englewood Cliffs, NJ: Prentice-Hall.
29. Faruque, M. A. A., Chhetri, S. R., et al. (2016). Acoustic side-channel attacks on additive manufacturing systems. In *International Conference on Cyber-Physical Systems (ICCPS).* Piscataway, NJ: IEEE.
30. Theodoridis, S., Pikrakis, A., Koutroumbas, K., & Cavouras, D. (2010). *Introduction to Pattern Recognition: A Matlab Approach: A Matlab Approach.* London: Academic Press.
31. Pedregosa, F., Varoquaux, G., Gramfort, A., Michel, V., Thirion, B., Grisel, O.,et al. (2011). Scikit-learn: Machine learning in Python. *Journal of Machine Learning Research, 12*, 2825–2830.
32. Printrbot 3D Printers. (2015). www.printrbot.com.
33. Zoom H6 Handy Recorder. (2015). www.zoom-na.com.
34. MATLAB. (2015). *(R2015b).* Natick, MA: The MathWorks.
35. Python 2.7.10. (2015). www.python.org.
36. Barry, D., Coyle, E., & Lawlor, B. (2004). Real-time sound source separation: Azimuth discrimination and resynthesis. In *Audio Engineering Society Convention* (Vol. 117). New York, NY: Audio Engineering Society.
37. Syskind, M. P., Larsen, J., Kjems, U., & Parra, L. C. (2007). A survey of convolutive blind source separation methods. In *Multichannel Speech Processing Handbook* (pp. 1065–1084).

Chapter 3
Aiding Data-Driven Attack Model with a Compiler Modification

3.1 Introduction

Cyber-physical systems (CPS) have been recognized as the enabler of the fourth industrial revolution (Industry 4.0) [1]. Due to the tight integration of the cyber and the physical world [2, 3], CPS allows us to enhance our interaction with the physical world around us. In fact, with CPS as the foundation, next generation of smart manufacturing is assumed to be more robust and intelligent. This forecast, however, does not consider the challenges [2] associated with a pervasive diffusion of CPS in every aspect of our lives, including the most sensitive and critical ones. One of the major challenges which designers have to address is to ensure the secure and reliable operation of these systems [4, 5].

One of the fundamental properties which security has to guarantee is *confidentiality*. A breach in confidentiality will cause significant economic damages. This is true when the lost information is, for instance, an intellectual property (IP). Overall, loosing IP will increase the financial loss in the manufacturing sector. The problem of guaranteeing *confidentiality* of the IP has been extensively studied by the research community, also related to the field of CPS [6, 7]. Interestingly, the confidentiality of IPs built using additive manufacturing, can be broken by exploiting side-channel information for reconstructing the G/M-Code used by 3D printers to generate the objects [8, 9]. G/M-codes are instructions sent to the 3D printer to control its motion and the machine parameters.

In the previous chapter, we presented a data-driven attack model that utilizes the acoustic side-channel to breach the confidentiality of the system. In this chapter, we introduce and demonstrate a novel and powerful attack to IPs fabricated with 3D printers. By altering the compiler used to generate the G/M-code, a malicious tool provider can generate codes which, while still producing the desired object, increases the amount of information leaked during the prototyping process. This significantly simplifies a reverse engineering attack carried out by exploiting the

© Springer Nature Switzerland AG 2020
S. R. Chhetri, M. A. Al Faruque, *Data-Driven Modeling
of Cyber-Physical Systems Using Side-Channel Analysis*,
https://doi.org/10.1007/978-3-030-37962-9_3

side-channels. To implement a modified compiler capable of aiding such an attack, we first analyze the information leakage from several side-channels (including acoustics, electromagnetic, power, and vibration), then we identify the strategies to maximize the leakage from each channel, and finally we demonstrate these strategies in the compilation of the G/M-codes. The approach followed in this chapter is motivated by attack models presented in the domain of hardware and software security. In these domains, hardware Trojans and malware have effectively utilized covert channels to leak information from the system. Authors in [10] inserted a malware in a system to convert the hard disk drives LED to leak sensitive data from air-gapped computers. Authors in [11] demonstrate how hardware Trojan can enhance the information leakage from the power side-channel.

The aim of this chapter is to demonstrate the feasibility of cross-domain attack model, where an attacker introduces change of cyber-domain variables to increase the amount of information leakage in the physical-domain. We believe that this type of attacks are not yet explored in depth by the scientific community, especially in the context of cyber-physical systems.

3.2 Attack Model Description

In previously proposed side-channel attacks on 3D printers [12–14], an attacker is motivated to steal the intellectual property (IP) inherent in the 3D object printed by the designer. The diagram of the attack is depicted in in Fig. 3.1. In this chapter, we envision a stronger attacker, having the same motivation, as the previous ones, but having also access to the tool chain used during 3D printing. The IP target of the attack, ultimately consists into the design file of the printed object, which is stolen by the adversary in the form of various parameters such as geometry, process parameters, machine parameters [8]. Since most of these 3D printers are used in prototyping stages, the attacker is especially motivated to steal the geometry information during the prototyping to reconstruct the object and gain a significant competitive advantage.

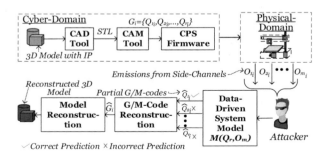

Fig. 3.1 Example of side-channel attack on cyber-physical manufacturing system [12–14]

Historically, in side-channel attacks, the adversary is assumed to have only low-level access (or be in the close proximity of to the system whereby he/she can non-intrusively monitor the side-channels by surreptitiously placing sensors such as microphones and accelerometers. Since side-channel attacks do not require alteration in the devices themselves, or do not require other invasive alteration such as access to the network, they could be harder to be identified and prevented. These attacks can be carried out by malicious employees having access to the location where printing is carried out, but incapable of having access to the design flies, to make an illegitimate copy. It has been shown that these types of side-channels offer can be successfully used to reconstruct information about the IP being printed by 3D printers used for prototyping [12–16].

To mount his attack, as is usually done in template attacks, the adversary will first train a model, using a device similar to the one which will be target of the attack, and then he/she will use the model to classify and reconstruct the code starting from the information leaked by the device under attack. During the training phase, the attacker will characterize the G/M-code by acquiring various side-channels (and extract various time and frequency domain features from them). These data will be used to train and create a model to link the G/M-code and the related analog emissions. As mentioned earlier, this process is done using a 3D printer similar to the target one. An attacker may create models also using advanced machine learning algorithms (such as random forests [17] or neural networks [18]) improving the accuracy of the model and, ultimately, improving the attack. The reconstructed G/M-codes are partial, as will be explained in Sect. 3.3. During an attack phase, when designers use the 3D printer, the adversary acquires these analog emissions and the previously trained classifier to reconstruct the partial G/M-codes, and eventually reconstruct the whole design file of the printed object. However, as shown in Fig. 3.1, these models are not capable of accurately predicting the partial G/M-codes. An attacker may therefore reconstruct G/M-codes that do not completely matches the printed 3D object.

To improve the reconstruction of partial G/M-codes, in this chapter we present a new attack methodology. Current side-channel adversaries are assumed to not have any access to the digital design process of the devices under attack, neither they are assumed to have access to the software compilation tool chain. However, this situation is rapidly changing, since stronger adversaries, capable of combining design alterations (such as hardware Trojans) with side-channel attacks are considered realistic and thus studied more in depth (previous works [19], for instance, demonstrated the possibility of increasing the side-channel leakage of a protected logic style by altering the doping polarity of logic gates). In this chapter, we consider an adversary who is capable of altering the design tools used to produce the file to be printed and combine this modification with side-channel attacks to recover the design file. We believe that threats coming from design tools are realistic, since they are gaining importance also in other fields of security [20, 21].

The new attack model that incorporates the compiler modification is shown in Fig. 3.2. The attacker is capable of tampering with the tool chain of the 3D printing. The infiltration may be in the form of a modified computer-aided design tool,

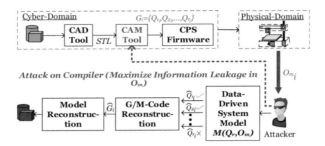

Fig. 3.2 Attack based on altered Compiler

computer-aided manufacturing tool, or the firmware of the 3D printer itself. Most of these desktop 3D printers used for prototyping use open source CAD, CAM, and firmware which are easily accessible by an attacker [22, 23]. The attacker only needs to modify and change these tools once, whereby they add malicious altered codes that increase the information leakage through the side-channels without hampering the normal operation of the 3D printer. In terms of accessibility, accessing the digital process chain is a challenge; however, this access is only required once, for instance, during the delivery of the device or of a software update.

The access needed for the first part of the attack can be possible in at least three scenarios. In the first one, the tool provider or the printer manufacturer introduces the malicious modifications in tool chain or in the firmware of the printer and the modification would be present in all the printing files. This threat is similar to the threat model for hardware Trojans previously mentioned, where a Trojan, inserted by a malicious foundry, alter the side-channel robustness of a chip by weakening the countermeasure. In the second scenario, a malicious employee, which has access to the source code of the fabrication tool chain or the firmware (most of them being open source for desktop 3D printers), maliciously modify them. Modification of the tool chain might be more appealing forms of cyber-attacks, than a simple copy of design files (for example, CAD design), since they are much stealthier and more difficult to detect. Thus, they could leave the possibility to reverse engineering designs for much longer time without the knowledge of the company. The third scenario is interdiction, which has been rumored to be used in the past [24–26] and have been recently proven to be practically feasible [27]. Interdiction could be used by competitors to get advantages or to damage other companies. In this scenario, during interdiction, a competitor can intercept the printing device during the delivery, modify the compiler or the firmware, and then proceed with the shipping to the customer. In this case, the main reason to increase the side-channel of the printing process instead of injecting other type of modifications is mainly due to the stealth of the attack and to the (almost) impossibility of demonstrating the malicious nature of the modifications.

In this chapter, to demonstrate the feasibility of attacks maximizing the information leakage through the side-channels, we have targeted the CAM tool (Sect. 3.3), which compiles the G/M-codes from the STL file. By incorporating altered codes, we will demonstrate that an attacker may improve their success rate for predicting the partial G/M-codes without hampering normal operation of the 3D printer. Let us take an example of reconstructing the partial G/M-code with and without using the attack on the compiler. While reconstructing the G/M-codes, one of the partial G/M-codes is the angle of a line segment that is moving in each of the layers. When the angle of the movement is $45°$ the line moves equally in both X and Y axes. This means that the actuators, such as stepper motors, responsible for moving the nozzle of the 3D printer have similar control signals. If their physical implementation (the load carried, the framework to hold the stepper motor, etc.) is the same then they will give off analog emissions, such as acoustics, that are quite similar. This will make differentiating the angle variation between the X and Y movements a non-trivial task. In such a scenario, an attacker may utilize an altered compiler to reveal the angle information by introducing code that triggers each time the angle changes. For example, it can be done by slightly changing the speed of the nozzle-cooling fan every time the angle changes without it being perceptible to the user. In the next section we will present our proposed compiler-based attack model in detail.

3.3 Compiler Attack

The overview of our attack is depicted in Fig. 3.3. The main objective of the adversary in our scenario is to recover the IP (the G/M-code) and other available design parameters. An attack aiming at reconstructing the G/M-code starting from

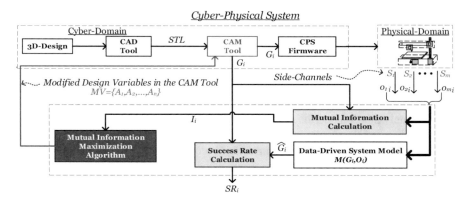

Fig. 3.3 The proposed compiler attack: the adversary identifies the variables controlling the 3D printers capable of maximizing the side-channel leakage, then produces an altered compiler capable of generating G/M-codes maximizing these leakages by modifying the use of these variables. When needed, the adversary recovers the IP produced by the 3D printer with an easier side-channel attack

side-channel analysis consists of two phase: a *profiling phase* and an *attack phase*. In the following part of this section we will explains these two phases in details.

3.3.1 Profiling Phase

In the *profiling phase*, an adversary is assumed to have access to the target device or to an identical replica of the target device. This profiling phase is similar to the profiling carried out during the template attacks [28], which have been proven by previous research as being an extremely powerful form of side-channel attack [28]. Using the n training G/M-codes, $G_n = [g_1, g_2, \ldots, g_n]$, the adversary can characterize m side-channels, $S_m = \{S_1, S_2, \ldots, S_m\}$. From each side-channel they are able to observe the analog emissions represented by random variables $O_m = \{O_1, O_2, \ldots, O_m\}$. We represent these emissions with random variables as we would like to increase the leakage by maximizing the mutual information between the emissions and design variables (see Sect. 3.3.3). An attacker collects data from each of these side-channels, filters them, and extracts several time and frequency domain features (let us say l features in total). For each of the side-channel an attacker eventually obtains the following matrix for each side-channel:

$$O_i = \begin{bmatrix} o_{i_1}^{f_1} & o_{i_1}^{f_2} & o_{i_1}^{f_3} & \ldots & o_{i_1}^{f_l} \\ o_{i_2}^{f_1} & o_{i_2}^{f_2} & o_{i_2}^{f_3} & \ldots & o_{i_2}^{f_l} \\ \vdots & \vdots & \vdots & \ddots & \vdots \\ o_{i_n}^{f_1} & o_{i_n}^{f_2} & o_{i_n}^{f_3} & \ldots & o_{i_n}^{f_l} \end{bmatrix} \quad i = 1, 2, 3, \ldots, m \tag{3.1}$$

where $o_{i_n}^{f_l}$ is the value of the random variable O_i for side-channel S_i, n is the total emission data collected, and l represents the total features extracted from the emission data. Although different number of features can be extracted from each of the side-channels, here we assume that similar and equal number of features are extracted. Depending on the attacker, they may use *uni-variate* or *multi-variate* analysis to estimate a model function $\hat{M}$ that approximates the relation between the side-channel emission and the cyber-data. We assume that an attacker has unlimited access to the system closely representing the target 3D printer, such that they can acquire large samples of analog emission O_i corresponding to the G/M-code G_i to accurately estimate the model functions using machine learning algorithms. As mentioned, these assumptions have proven to be realistic by the amount of literature dealing with template attacks. The model function can simply be chosen as a function with k values which satisfies the following criteria:

$$k = \arg\min_k \sum_{i=1}^{N} \left| g_i - \hat{f}_k(o_i, .) \right| \tag{3.2}$$

where $k = 1, 2, \ldots, N$ is the number of various model functions (such as neural networks, random forests, support-vector machines, etc.) used to estimate the model function. The model function is trained to estimate the G-code, $\hat{g}_i = \hat{f}_k(o_i, \phi)$, given the corresponding analog emissions. $\phi = \{\phi_1, \phi_2, \phi_3, \ldots, \phi_t\}$ is a vector of parameters used to tune the model function. Corresponding to the model function used (indexed by k in Eq. (3.2)), length t and value of ϕ is determined.

We carried out the profiling phase in a relatively protected environment, keeping the printer in a room which was not accessed during the whole phase. However, the room was separated from the rest of the environment only by a closed door, thus possible noise coming from outside could have been included in the model. The printer was placed on a regular table, with no support for attenuation of vibrations. Power and electromagnetic measurements have also been carried out on the regular device operating in a non-insulated environment. We believe that a profiling carried out in a much more insulated environment would lead to better results, since it would place the attacker into a more ideal situation. However, since we are interested in the real impact of the proposed attack, we constructed our setup in a way to mimic the setup which an adversary could easily mount using relatively inexpensive equipment. As in any template attack, the closer is the profiling setup with the target one, the higher are the chances of success. However, the attack we propose here aims to amplify the leakage of information, thus making easier to be detected by an adversary also when the profiling setup is very different from the attack one.

3.3.2 Attack Phase

During the *attack phase*, the model function $m : G \rightarrow O$ or $\hat{f}_k(., \phi)$ acquired from Eq. (3.2), is used to reconstruct the G/M-code. The model function is estimated using various algorithms (such as neural networks, random forests, support-vector machines, etc.). For each of these algorithms k, we select the function that gives the least analog emission to G-code estimation error. An attacker has to acquire all the analog signals from the side-channels to estimate the complete G/M-code. The accuracy of the estimated function determines the accuracy of the attack. In most of the attacks, the reconstruction of the G/M-code is done using multiple model functions, each responsible for estimating partial information about the G/M-code. Let the G/M-code be partitioned into $\{Q_1, Q_2, Q_3, \ldots, Q_r\}$ of disjoint blocks, such that $\cup_{i=1}^{r} Q_i = G$. This partition is based on the fact the each instruction of G/M-code is responsible for accomplishing multiple actuation at the same time (for example determining angle and length of the line segment being printed). Hence, we can decompose such G/M-code further into smaller control actions. Then we have model functions $\{\hat{m}_1(Q_{1_i}, O_i), \hat{m}_2(Q_{2_i}, O_i), \ldots, \hat{m}_r(Q_{r_i}, O_i)\}$ that collectively estimate the G/M-code. This is especially true in additive manufacturing, because the IP is hidden in the geometry of an object, and the main information that reveals the geometry of the object while printing are: travel *feed-rate* of the nozzle, *angle* of the line segment, *distance* movement in each axis, etc. This information is present in G/M-code along with other information. Hence, only subset of $\{Q_1, Q_2, Q_3, \ldots, Q_r\}$ maybe required to reconstruct the geometry. Also, different side-channels reveal different sets of information about the G/M-code.

However, reconstructing the G/M-code can be problematic considering that the attacker, in a real scenario, might not have access to an environment where it is possible to collect noise-free samples of each side-channel emission. To improve the situation, we envision a stronger attacker, which is capable of modifying and producing G/M-code via compiler to maximize the leakage of information via side-channels. In fact, by modifying computer-aided manufacturing tools (such as *slicing algorithm* and *tool-path generation* algorithm), an attacker can easily introduce variation in the design variables $\{\alpha_1, \alpha_2, \alpha_3, \ldots, \alpha_n\}$, such that the estimated model functions $\hat{f}_k(., \phi)$ can more accurately reconstruct the G/M-code.

3.3.3 Compiler Modification

To maliciously modify the compiler to generate G/M-code with higher information leakage, we need to identify the variables capable of maximizing the leakage through the printer components. These variables are machine-specific. In this work, we identified three components which can be easily used to maximize information leakage. Our malicious compiler will utilize these variables without affecting the normal operation of the 3D printer. In detail, the variables which we used are:

1. **Fan Speed** (A_1)**:** The machine code $M106$ in the 3D printer is responsible for controlling the fan speed. This fan is used to control the temperature of the nozzle. The speed of the fan can vary from 0 to 255. A minute changes in the fan speed is not noticeable to human perception; however, it can affect the acoustic, magnetic, power, and vibration side-channels. These changes in the fan speed must not cause the temperature of the nozzle to negatively impact the smooth flow of materials. This has to be considered in the altered compiler.

2. **Speed at the End of Line Segment** (A_2)**:** In order to reverse engineer the 3D object, an attacker needs to reconstruct each line segment that is used in printing the 3D object layer by layer. Hence, by varying the speed at the end of the line segment, an attacker can increase the speed variation drastically between two line segments, allowing them to decipher the change of the line. However, the speed variation cannot be large to cause mechanical shock or hamper the deposition of the material through the nozzle.

3. **Power to Stepper Motors** (A_3)**:** M-codes $M17$ powers all the stepper motors, whereas $M18$ disables all the stepper motors. This variable can be used to enhance the information leakage in the power and electromagnetic side-channels. However, the stepper motors should be turned off and then on quickly so as not to affect the operation while affecting the side-channels enough to reveal certain information.

There are several other G/M-codes that may be utilized by the malicious compiler to optimally leak information from the side-channels without the knowledge of the legitimate user such as M108 to set extruder speed, M113 to set extruder's Pulse Width Modulation (PWM), M140 to set printing bed temperature, M204 to set the acceleration of stepper motors, etc.). Here, we only present few for demonstrating the effectiveness of our attack model.

To quantify the amount of information leaked from the side-channels, we use as metric the *mutual information* between the partitioned G/M-code Q_r and the observed analog emission in each of the side-channels. Intuitively, the mutual information indicates how much information of a specific G/M-code Q_r is available if the analog emission O_m is known. We denote the discrete random variable with the capital letter O, Q, G and the values taken by these random variables with the small letters o, g, q. Although the value of analog emissions from the side-channels maybe calculated using the first principle based equations describing the physical characteristics of the system, this step is non-trivial. Instead, we represent such analog emission as random variables and observe its realization using the sensors. Moreover, depending on the 3D object being printed, slicing algorithms, process parameters, etc., different G/M-code can be passed to the 3D printer. Since we do not assume an attacker to know the 3D object or process parameters beforehand, we treat these G/M-code being passed to the 3D printer as random variables with some underlying distribution. The mutual information is then defined as [29]:

$$I\left(Q_r; O_m\right) = \sum_{o_m \in O_m} \sum_{q_r \in Q_r} p\left(q_r, o_m\right) log_2 \left(\frac{p(q_r, o_m)}{p(q_r)p(o_m)}\right) \qquad (3.3)$$

where $p(q_r, o_m)$ is the *joint probability distribution function*, and $p(q_r)$ and $p(o_m)$ marginal probability distribution for the random variables Q_r and O_m. The joint probability function may be calculated using training G/M-codes consisting of range of uniformly distributed values for Q_r and collecting the corresponding analog emissions from the side-channels. Since we have used base two for the logarithm, the unit of the mutual information is *bits*.

3.3.4   Transformations for Leakage Maximization

As mentioned earlier, the proposed attack consists of two phases: first, a profiling phase and second, an attack phase. The profiling phase is carried out only once, and it is done for two reasons. The first one, is to understand how and where to inject the malicious modification in the compiler (or in the firmware of the printer), the second is to build the templates which will be used to reconstruct the code during the second phase of the attack. During the profiling phase, the attacker estimates the model function. Based on the model function, the mutual information between the analog signals and the G/M-code is computed along with the accuracy of the reconstructed

G/M-code. Then, various design variables are explored, and the ones that could maximize the mutual information for a given side-channel without affecting the normal operation of the 3D printer are modified accordingly. The altered compiler is then inserted into the design chain of the additive manufacturing system to produce G/M-codes that can be easily reconstructed by an adversary.

The goal of the attacker is maximizing the side-channel information. To do so in a systematic way, we develop a mutual information maximization algorithm. Our algorithm takes the current mutual information value between the partial G/M-code Q_r and the analog emissions O and maximizes the information available.

Algorithm 1: Mutual Information Maximization Algorithm

Input: Design Variables $\{A_1, A_2, \ldots, A_n\}$
Output: Design Variables Values $\{\alpha_{m_1}, \alpha_{m_2}, \ldots, \alpha_{m_n}\}$
1 **for** $i = 1 : n$ // Offline Data Collection
2 **do**
3   Define step size $\triangle_{A_i}$ and range min_{A_i} and max_{A_i}
4   **for** $k = min_{A_i} : \triangle_{A_i} : max_{A_i}$ **do**
5   **for** $j = 1 : r$ **do**
6   Define step size $\triangle_{q_j}$ and range min_{q_j} and max_{q_j}
7   **for** $l = min_{q_j} : \triangle_{q_j} : max_{q_j}$ **do**
8   Acquire analog emissions $o_{m_{(l.j.i)}}$ from m side-channels
9   Estimate Joint Probability Distribution Functions $p_{Q_r, O_m | A_i}(q_r, o_m | \alpha_{m_i})$
10   Calculate $I_{m_{i_k}} = I(Q_r, O_m)$ for each value of k
11 Estimate Non-linear function $\hat{f}_{m_n}(.)$ // Based on $I_{m_{i_k}}$
12 **for** $i = 1 : m$ **do**
13   **for** $j = 1 : n$ **do**
14   Optimize arg $\max_{(\alpha_{i_j})} \hat{f}_{i_j}(.)$

15 **return** $\{\alpha_{m_1}, \alpha_{m_2}, \ldots, \alpha_{m_n}\}$

Algorithm 1 performs the mutual information maximization. The input to the algorithm are the considered design variables $\{A_1, A_2, \ldots, A_n\}$. For each of the design variables, the algorithm returns values $\{\alpha_{m_1}, \alpha_{m_2}, \ldots, \alpha_{m_n}\}$ that maximize the amount of mutual information in each of the m side-channels. From lines 1–8, the algorithm collects the data for estimating the joint probability distribution function in line 9. Using this joint probability function, along with the entropy of the analog emissions and the pre-determined partitioned G/M-codes Q_r used for profiling, the conditional mutual information based on the value of each design variable is calculated in line 10. Then a non-linear function $\hat{f}_{m_n}(.)$ is used to estimate the relation between the mutual information and the values of the design variables in line 11. Then for each of the design variables and the side-channels, an optimization algorithm is used that maximizes the amount of the mutual information in each of the side-channels. The constraints for the optimization algorithm are: (1) maintain the printing time $T_p \leq T_I + \triangle_T$, where T_I is original printing time and $\triangle_T$ is the allowed threshold for time variation, (2) avoid large variation in nozzle temperature $\theta_I - \triangle_{\theta_L} \leq \theta_p \leq \theta_I + \triangle_{\theta_H}$, where θ_I is original temperature and $\triangle_{\theta_L}$, $\triangle_{\theta_H}$ is the allowed lower and upper threshold for temperature variation, (3) avoid large speed fluctuation $\vartheta_I - \triangle_\vartheta \leq \vartheta_p \leq \vartheta_I + \triangle_\vartheta$, where ϑ_I is original printing speed and $\triangle_\vartheta$ is the allowed threshold for speed variation, (4) and not keep stepper motors idle for a longer than period of time $T_{IDLE} \leq \triangle_{IDLE}$, where $\triangle_{IDLE}$ is the allowed idle time without causing the stepper motors to move freely. To obtain these thresholds, an attacker first needs to find the optimal value of these parameters θ_p, ϑ_p, T_{IDLE}, and T_I. These parameters can easily be accessed if the attacker has a profiling 3D printer which is similar to the target 3D printer. For some of the parameters, the optimal range of the values are given in the user manual. However, an attacker may go beyond the optimal range by increasing or decreasing the threshold and analyzing its corresponding effects on quality of the 3D objects.

The second phase of the attack is a classical template side-channel attack. To carry out that, the adversary need to have physical access to the device during the printing process. The type of access required depends on the particular "channel" that the adversary wants to exploit. For power, a complex and likely to be visible measurements setup maybe required. For acoustic, vibration, and EM side-channel attacks, simple modern smart phones [12] have been demonstrated to have the potential to be sufficient to reconstruct the whole G/M-code. After the channel traces are collected, the adversary will attempt to reconstruct the G/M-code using a model function estimated during the profiling phase.

3.4 Experimental Results

Our experimental setup is shown in Fig. 3.4. We have used a Cartesian desktop 3D printer (*Printrbot*) which uses fused deposition modeling technique for printing. In order to monitor the analog emissions from the side-channels we have used three acoustic sensors (*Audio-Technica AT2021*), AC/DC current clamp (*PICO TA018*),

Fig. 3.4 Experimental setup for the proposed attack

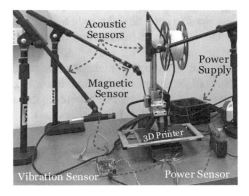

magnetic field sensor (*HMC5883L*), and accelerometer (*uxcell ADXL335*). The acoustic sensors are placed in X, Y, and Z-axes perpendicular to each other, the vibration sensor is placed at the base of the 3D printer, magnetic sensor is placed on the table close to the 3D printer, and the power sensor is clamped in power cable between the power supply and the 3D printer. Data acquisition from all the side-channels is done using National Instruments data acquisition hardware (*NI USB-6229*). The analog emissions collected from the side-channels are then processed in Matlab (*R2015b*). To measure the change in the mutual information caused by changing various parameters, we have created line segment with multiple layers and angle to the X axis. For calculating the success rate, we have used polygons with various sides and height without infill as our test objects. We have implemented the first and second phase of the proposed attack model in the same 3D printer.

We partition the G/M-codes into three sets. An attacker capable of reconstructing these subsets of G/M-codes accurately is able to reconstruct any line segment being printed by the 3D printer at any instance of time.

1. **Angle of the Line Segment** (Q_1): Each instruction of the G/M-code moves the nozzle of the 3D printer from point of origin to the point of destination making a line segment. Since, an object is printed layer wise, most of these line segments lie in the XY-plane. Hence, we define the angle made by the line segment with the X axis as one of the variables of the G/M-code.
2. **Change of Layer** (Q_2): When the 3D printer is done printing on one layer, it moves to another layer. Each G/M-code consists of information on which layer it is printing on. Hence, we define Q_2 as the variable that determines the change of layer.
3. **Change of Line Segment** (Q_3): The change of line segment is not directly derived from the G/M-code, but rather from the analog emissions acquired when a G/M-code instruction begins and ends. We specify this variable as Q_3.

These three variables are not an exhaustive partitions of the G/M-code. However, they are enough to trace the nozzle movement, and determine the geometry of the 3D-object being printed, as in conjunction, these variables define the vector of line segments that make up each layer, define various layers, and finally assign the analog emissions to specific G/M-code instructions.

The leakage of information can be arbitrary increased. However, we need to ensure that while increasing the leakage of information, we do not make the alteration of compiler understandable. To do so, we empirically determined the values of threshold which would allow to increase the leakage of information while still remain unnoticed. For providing the constraints to the optimization algorithm, we selected the threshold values as follows: (1) $\triangle_T = 1\%$ of the T_I, we assume that 1% printing time is not noticeable for any object being printed. (2) $\triangle_{\theta_L} = 10°$ and $\triangle_{\theta_H} = 5°$. The thermoplastic extruded in the given 3D printer is a PolyLactic Acid (PLA), which is a bio-polymer with melting temperature between 180° to 220° [30]. The original printing temperature for given 3D printer is normally set at 205°. Hence, this gives the compiler some room for changing the fan speed without compromising the printing quality. (3) $\triangle_\vartheta = 10\%$ of the normal printing speed, which is not large enough to cause jerky motions but may cause the analog emissions to be noticeable in the side-channels. (4) we assigned $\triangle_{IDLE} \leq 30\,\text{ms}$ for minimizing the free movement of stepper motors. The compiler alteration attack is compared with the attack methodologies incorporated in [12–14], where side-channel data are collected and models are trained to predict various G/M-code information.

3.4.1 Accuracy Metric

A better metric to evaluate the effectiveness of our attack is success rate of an attacker. Intuitively, the success rate measures the capability of an adversary to recover the G/M-code (or part of it with certain error threshold) given a specific leakage and, contrary to the mutual information, is dependent on the specific attack scenario. Let us assume that any side-channel $S' \in \{S_1, S_2, \ldots, S_m\}$ leaks a subset of information $Q' \subseteq Q$ with $Q' = \{Q_1, Q_2, \ldots, Q_t\}$, where $t \leq r$ and $t \in \mathbb{Z}_{>0}$, with r being possible number of disjoint sets into which the G/M-code can be partitioned into. We assume that this subset of information is able to explain the geometry of the 3D object. Let us define a variable e_i, such that $e_i = 1$ when $|q_{r_i} - \hat{m}_r(o_{m_i}, .)| \leq \triangle_e$, and 0 otherwise, where $\triangle_e \in \mathbb{R}$ is the error threshold for which the estimated partial G/M-code can be used to reconstruct the complete G/M-code. Ideally, the error threshold for reconstruction should be close to the process variation for an attacker to have full reconstruction capability. However, to demonstrate that the success rate can be improved by utilizing the subtle compiler alteration, the error threshold is fixed with reference to the accuracy of one of the most powerful side-channel for the given 3D printer (determined experimentally). Let us say from the experiment we obtain four error thresholds $\triangle_{e_a}, \triangle_{e_v}, \triangle_{e_m}$, and $\triangle_{e_p}$ for acoustic, vibration, EM, and power side-channels, respectively, for each partial G/M-codes, that gives us more than 75% success rate in reconstruction. Then we chose the smallest $\triangle_e$, for calculating the success rate for all the side-channels as well. For each model function $\hat{m}_r(.)$, estimated for the side-channel S', we define partial success rate in estimating a portion of G/M-code Q'_t as follows:

$$SR_{(t,m)} = \frac{\sum_{i=1}^{N} e_i}{N} \qquad (3.4)$$

where N gives the total number of G/M-code instructions used in describing the 3D model. However, in calculating the total success rate, the consideration is given for accurate estimation of all the partial G/M-code. Hence, an attacker is successful in reconstructing the G/M-code if and only if $e_{i_1} = 1, e_{i_2} = 1, \ldots, e_{i_t} = 1$, where e_{i_j} is the error while estimating j_{th} partial G/M-code for i_{th} instruction. With this the total success rate for the given side-channel S' can be written as follows:

$$SR_{(m)} = \frac{\sum_{i=1}^{N} \prod_{j=1}^{t} e_{ij}}{N} \qquad (3.5)$$

For each of the side-channels and the partitioned G/M-code, we measure the amount of information leaked and the corresponding success rate of the estimated model functions $\hat{m}_r(.)$. Then, we incorporate our mutual information maximization algorithm to modify the design variables and see the corresponding changes in the success rate of the same estimated model function. In quantifying the leakage, we extract power spectral density as a feature over the range of 0 Hz–10 KHz evenly distributed at an interval 20 Hz due to the fact that the emissions are with in this range during initial spectral analysis, and 20 Hz interval resolution is able to capture the variation in the frequency domain), perform Principal Component Analysis (PCA) on the feature set this will make it possible to transform the correlated frequency components into uncorrelated principle components, and reduce the dimension for the attack model), and present the mutual information of the highest principle component. In order to measure the mutual information for each of the partitioned G/M-code $\{Q_1, Q_2, Q_3\}$, we need to calculate the corresponding analog emissions signals. Hence, in our experiment, for the angle (Q_1), we have varied the angle from $0°$ to $360°$ at an step angle of $9°$. We, assume Q_1 is a random variable with uniform distribution. Hence, its total entropy will be $log_2(40) \simeq 5.32$ bits. Random variable Q_2 and Q_3 only have two possible values, whether the layer changed or not, and whether it is the same line segment or a different one. Hence, their entropy is $log_2(2) = 1$ bits. For demonstration purpose, the training data consisted of line segments varying in angle, layer (10 layers) and direction (by printing line segments in opposite directions).

3.4.2 Mutual Information

The mutual information is calculated for each partial G/M-code $\{Q_1, Q_2, Q_3\}$ (shown in Figs. 3.5, 3.6 and 3.7) before and after altering the compiler (the slicing and the tool-path generation tool). These values are calculated based on the joint probability distribution estimated by assuming a uniform distribution for partial G/M-codes, and collecting the corresponding analog emissions from the side-channels. In Fig. 3.5, the mutual information before and after the alteration of the compiler is shown between each of the side-channels and the partial G/M-code Q_1. It can be observed that the power side-channel, even though had lower overall mutual information compared to other side-channels, has highest increase (**+21.5%**) in mutual information after altering the compiler. Vibration side-channel has highest mutual information with the partial G/M-code. This means that more information about Q_1 can be inferred from the vibration side-channel compared to other side-channels.

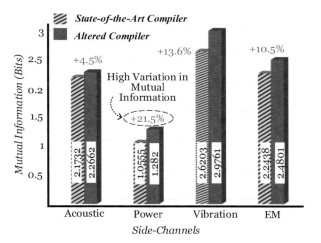

Fig. 3.5 Mutual Information between Q_1 and Side-channels before [12–14] and after altering the Compiler

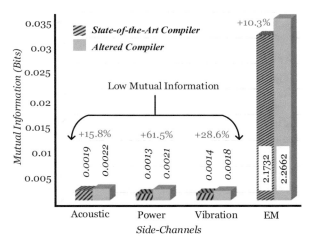

Fig. 3.6 Mutual Information between Q_2 and Side-channels before [12–14] and after altering the Compiler

In Fig. 3.6, the mutual information before and after the alteration of the compiler is shown between each of the side-channels and the partial G/M-code Q_2. It can be observed that the mutual information is highest for electromagnetic side-channel, although increase (**+10.3%**) in the mutual information after the alteration of the compiler is the lowest for it, whereas power side-channel shows large increase in mutual information after the alteration of the compiler (**+61.5%**).

In Fig. 3.7, the mutual information before and after the alteration of the compiler is shown between each of the side-channels and the partial G/M-code Q_3. For Q_3, similar to Q_2, electromagnetic side-channel has the highest mutual information with increase of mutual information by **+26.3%** after the compiler alteration. Power side-channel, however, showed significant increase in mutual information. However,

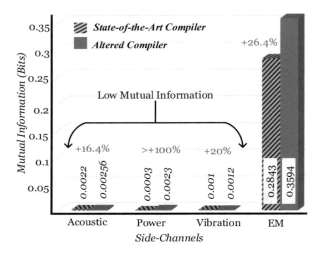

Fig. 3.7 Mutual Information between Q_3 and Side-channels before [12–14] and after altering the Compiler

compared to the electromagnetic side-channel it has relatively very low mutual information.

We can see that for the partitioned G/M-code Q_1, the vibration side-channel has most mutual information, whereas the power side-channel has the least. This is due to the fact that angle variation does not affect the constant DC power supply. Compared to the attack models used in [12–14], the altered compiler increases the mutual information in each of the side-channel. The increase in mutual information for all the side-channels demonstrates that the alteration of the compiler, while does not change the final design, helps attackers to infer original G/M-code more accurately. Contrary to other forms of breaching the confidentiality by accessing the design files, these modifications allow an attacker to obtain the G/M-code in a way which is extremely hard to be detected.

3.4.3 Partial Success Rate

Before calculating the success rate, the data collected using uniformly distributed partial G/M-codes is split into 3:1 training and testing set, respectively. Then *random forest regression* algorithm is trained to predict the angle from analog emissions collected from the side-channels, and *random forest classification* algorithm is used to predict the layer and line segments changes. Since only the angle is a continuous value, $\triangle_e$ is set as 3° based on the vibration side-channel, which provide more than 75% accuracy for prediction with error of ± 3°. Then the average partial success rate is calculated for individual side-channels and the partitioned G/M-codes. The results for partial success rate for reconstructing Q_1, Q_2, and Q_3 are presented in Figs. 3.8, 3.9, and 3.10, respectively.

In Fig. 3.8, the partial success rate before and after the alteration of the compiler is shown for the partial G/M-code Q_1. It can be observed that the vibration side-channel has the highest success rate for reconstructing Q_1, while only having **+6.53%** increase in the success rate after alteration of the compiler. Whereas, power side-channel has highest increase in success rate by **+75.77%**, while having the lowest overall success rate for predicting Q_1.

In Fig. 3.9, the partial success rate before and after the alteration of the compiler is shown for the partial G/M-code Q_2. It can be observed that the electromagnetic side-channel has the highest overall success rate. However, the success rate decreased by **0.17%**. This may be due to the fact the success rate is already higher, and change in the design variable does not improve the success rate drastically. The other reason for this could be that, although, the mutual information increased for Q_2, it only gives the upper bound for inferring about the random variable Q_2 given the analog emissions, and the success rate highly depends on the accuracy of the estimation function ($\hat{f}_k(., \phi)$). And in this experiment, the estimation function is

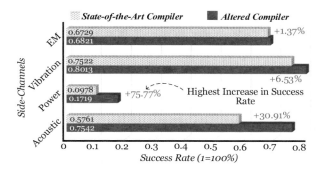

Fig. 3.8 Partial Success Rate for reconstructing Q_1 before [12–14] and after altering the Compiler

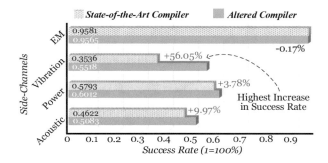

Fig. 3.9 Partial Success Rate for reconstructing Q_2 before [12–14] and after altering the Compiler

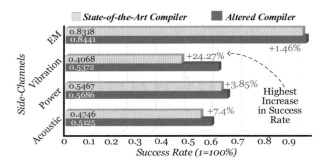

Fig. 3.10 Partial Success Rate for reconstructing Q_3 before [12–14] and after altering the Compiler

not able to utilize the increase in the mutual information. However, vibration side-channel showed drastic increase in the success rate by **56.05%** corresponding to the increase in the mutual information.

In Fig. 3.10, the partial success rate before and after the alteration of the compiler is shown for the partial G/M-code Q_3. Similar to Q_2, electromagnetic side-channel gave highest success rate for reconstructing Q_3. However, increase in success rate after the alteration of the compiler is only **1.46%** compared to vibration side-channel, where increase in success rate (**24.27%**) is the highest.

3.4.4 Total Success Rate

For calculating the total success rate, we have printed various regular polygons with number of sides ranging from 3 to 12, and measured the accuracy for reconstructing the partial G/M-code information (Q_1, Q_2, and Q_3). For each polygon, five layers are printed and the side length are varied from 5 to 20 mm. Then the prediction is performed based on the model function trained to calculate the partial success rates. Then based on these predictions, Eq. 3.5 is used to calculate the total success rate before and after the compiler is altered. From Fig. 3.11, we can observe that the total success rate after altering the compiler has increased for all the side-channels. The total success rate is highest for vibration side-channel around **24.27%**. However, it has the lowest increase in success rate compared to other side-channels. The power side-channel has the lowest overall success rate; however, it has the highest increase in the success rate (**39.71%**). Our attack model leaves a large margin to an attacker for finding an optimal combination of mutual information maximization across various side-channels for obtaining increase in success rate, while achieving higher stealth. Hence, combined with the altered compiler, attack models presented in [12–14] can achieve higher reconstruction accuracy for the G/M-code using the proposed attack model.

Fig. 3.11 Total success rate before [12–14] and after (proposed attack) altering the Compiler

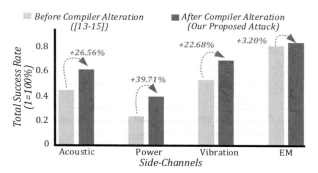

3.5 Discussion

Complete Object Reconstruction Here, a proof of concept work is presented for a new and much strong attack model, where an attacker may be able to perform one-time access to alter the compiler producing the printing file (or the firmware interpreting it) and, thanks to this, the strengths of side-channel attacks aiming at stealing intellectual property is increased. Contrary to other forms of information leakage, the ones released using side-channels are much more difficult to be identified (and thus stopped). We have demonstrated this by using an altered compiler, the success rate for predicting the objects made of line segments has been improved significantly. However, in order to achieve accurate reconstruction of the 3D objects with arcs, we will incorporate further design variable analysis that an attacker might use. Nonetheless, not all movements are non-linear, and this method will still give partial information about various line segments in the 3D object. Furthermore, we have only presented the analysis on FDM based desktop 3D printers. Higher-end FDM 3D printers may utilize better techniques to reduce the acoustic noise and vibration. Hence, further analysis is required to implement this attack in higher-end 3D printers.

Difficulty of Compiler Alteration Any form of alteration on the compiler have to be surreptitiously placed to avoid detection. Patched Win-RAR vulnerability have been previously used to manipulate the design files [31], and compromise the physical structural integrity from the cyber-domain. We assume that an attacker would be interested to steal the IP of the 3D object during the prototyping stage. Currently, there are multiple third-party compiler software that is able to generate the G/M-codes. Although, the desktop 3D printers may not be connected to the internet, the PC that have this software have internet connection. Hence, for altering the compiler an attacker would require both proficiency in hacking skills as well as additive manufacturing process chain knowledge, to introduce subtle changes that would not be noticeable by the user [31].

Alteration of CAD and Firmware We have demonstrated an attack where the compiler of the 3D printer is altered. However, as mentioned in the chapter, the principle we propose is valid also for an attacker who wants to alter the firmware or the CAD tool itself to increase the information leaked via side-channel. Hence, in our future work, we will complete our demonstration by altering also the firmware and CAD tools to increase information leakage in the side-channels.

Combination of Multiple Channels In this chapter, we considered an attacker which has access to one side-channel at the time (or, at least, aims at exploiting one side-channel at a time) and aims at maximizing it using an altered compiler. This is a first step into this direction. A future interesting research direction is the to explore the combination of multiple side-channel attacks (and the joint increase of side-channel leakage). We leave a systematic exploration of this aspect as a future work.

Sensor Position Since an attacker relies on the physical emissions from the side-channels, the position and number of sensors placed may affect the success rate of an attacker. To this regard, an attacker may require to explore multiple sensor positions and number of sensors to improve their success rate. In our experiment, we have performed primitive sensor position by placing the sensors in close proximity from the components responsible for emitting physical emissions. However, more extensive methods need to be explored to figure out the optimal number and position of sensors. We leave this for our future work.

3.5.1 Countermeasures

The proposed attack causes the loss of information using side-channel attacks, thus, as other side-channel, could be very hard to be detected, since legitimate users could not be able to perceive any change in the normal operation of the 3D printer (the final object remains unaltered and the printing time also increases by an imperceptible amount). Counteracting the attack proposed in this chapter can be done in two ways. In the first approach, the traditional formal verification methods needs to be used to check the integrity of the G/M-code to make sure that the 3D printer firmware or CAD and STL files are not compromised. Second approach would be to have a golden reference of an unaltered CAD, CAM, and 3D printer firmware to check any anomalous behavior in the side-channels. However, this approach requires us to have a golden model (which, if existing, could be used directly in the printing phase).

Regarding counteracting side-channel attacks in general, it is a very hard task. Contrary to what is done with side-channel attack such as power analysis or timing analysis, where most of the countermeasure are preventive (namely they aim at avoiding an attack), in the context of the 3D printer this approach does not seem to be optimal. For instance, hiding the side-channel signal would requires either to generate a side-channel signal which is constant (however, it will be very hard to guarantee that each shape is printed using the same amount of time, generating the same noise, producing the same vibrations) or to cover it creating artificial noise to complicate the reconstruction phase. A much better approach seems to be the one of detecting the attack and react to it only when identified. However, this approach is relatively new and requires to be explored much more in depth.

3.6 Summary

In this chapter, we have presented a proof of concept for a new attack consisting of a modification of the compiler to generate G/M-codes which allows an easier reconstruction of the printing files using side-channel attacks. We demonstrated how multiple design variables can be used to increase the success rate of the G/M-code reconstruction, and we implement a malicious compiler which demonstrate the feasibility of our attack, capable of achieving an increase up to 39% of success rate measured across acoustic, power, vibration, and electromagnetic side-channels compared to the traditional attack models [12–14] that rely only on snooping the side-channels. Our work demonstrates the real feasibility of attacks carried out by using altered tool chains which can cause large financial loss to manufacturing industry. Such an attack can be carried out in several scenarios, including the so called interdiction, where deliveries are intercepted an manipulated introducing back doors. Future works will explore the role of other variables that control parameters of 3D printers as well as the possibility of fusing multiple side-channels to improve the G/M-code recovery capability.

References

1. Lee, J., et al. (2015). A cyber-physical systems architecture for industry 4.0-based manufacturing systems. *Manufacturing Letters, 3*, 18–23.
2. Lee, E. A. (2008). Cyber physical systems: Design challenges. In *2008 11th IEEE international symposium on object oriented real-time distributed computing (ISORC)*. Piscataway: IEEE.
3. Chhetri, S. R., Wan, J., & Al Faruque, M. A. (2017). Cross-domain security of cyber-physical systems. In *2017 22nd Asia and South Pacific design automation conference (ASP-DAC)*. Piscataway: IEEE.
4. Cardenas, A., et al. (2009). Challenges for securing cyber physical systems. In *Workshop on future directions in cyber-physical systems security*.
5. Wu, G., et al. (2016). A survey on the security of cyber-physical systems. *Control Theory and Technology, 14*(1), 2–10.
6. Zeltmann, S. E., Gupta, N., et al. (2016). Manufacturing and security challenges in 3D printing. *Jom, 68*(7), 1872–1881.
7. Chhetri, S. R., & Al Faruque, M. A. (2017). Side-channels of cyber-physical systems: Case study in additive manufacturing. *IEEE Design & Test, 34*(4), 18–25.
8. Yampolskiy, M., Andel, T. R., et al. (2014). Intellectual property protection in additive layer manufacturing: Requirements for secure outsourcing. In *Proceedings of the 4th Program Protection and Reverse Engineering Workshop*. New York: ACM.

9. Sturm, L., Williams, C., Camelio, J., et al. (2014). Cyber-physical vulnerabilities in additive manufacturing systems. *Context, 7*(2014), 8.
10. Guri, M., Zadov, B., Atias, E., & Elovici, Y. (2017). LED-it-go: Leaking (a lot of) data from air-gapped computers via the (small) hard drive led. arXiv preprint:1702.06715.
11. Lin, L., Kasper, M., Güneysu, T., Paar, C., & Burleson, W. (2009). Trojan side-channels: Lightweight hardware Trojans through side-channel engineering. In *CHES* (Vol. 5747, pp. 382–395). Cham: Springer.
12. Hojjati, A., Adhikari, A., et al. (2016). Leave your phone at the door: Side channels that reveal factory floor secrets. In *Proceedings of the 2016 ACM SIGSAC Conference on Computer and Communications Security*. New York: ACM.
13. Song, C., Lin, F., et al. (2016). My smartphone knows what you print: Exploring smartphone-based side-channel attacks against 3D printers. In *Proceedings of the 2016 ACM SIGSAC Conference on Computer and Communications Security*. New York: ACM.
14. Al Faruque, M. A., Chhetri, S. R., et al. (2016). Acoustic side-channel attacks on additive manufacturing systems. In *International Conference on Cyber-Physical Systems (ICCPS)*. Piscataway: IEEE.
15. Chhetri, S. R. (2016). *Novel side-channel attack model for cyber-physical additive manufacturing systems*. http://escholarship.org/uc/item/6c52g94w.
16. Al Faruque, M.A., Chhetri, S. R., Faezi, S., & Canedo, A. (2016). *Forensics of thermal side-channel in additive manufacturing systems-semantic scholar*. Irvine, CA.
17. Liaw, A., & Wiener, M. (2002). Classification and regression by randomForest. *R News, 2*(3), 18–22.
18. Haykin, S., & Network, N. (2004). A comprehensive foundation. *Neural Networks, 2*(2004), 41.
19. Becker, G. T., Regazzoni, F., Paar, C., & Burleson, W. P. (2013). Stealthy dopant-level hardware Trojans. In *International Workshop on Cryptographic Hardware and Embedded Systems* (pp. 197–214). Berlin: Springer.
20. Potkonjak, M. (2010). Synthesis of trustable ICs using untrusted CAD tools. In *2010 47th ACM/IEEE Design Automation Conference (DAC)* (pp. 633–634). Piscataway: IEEE.
21. Polian, I., Becker, G., & Regazzoni, F. (2016). Trojans in early design steps—An emerging threat. In *TRUDEVICE - 6th conference on trustworthy manufacturing and utilization of secure devices*. http://hdl.handle.net/2117/99414.
22. Ultimaker. (2017). *Cura software*. https://ultimaker.com/en/products/cura-software.
23. Slic3r. (2017). http://slic3r.org/.
24. SPIEGEL Staff. (2013). *Inside TAO: Documents reveal top NSA hacking unit*. http://www.spiegel.de/international/world/the-nsa-uses-powerful-toolbox-in-effort-to-spy-on-global-networks-a-940969.html.
25. Macri, G. (2014). *Leaked photos show NSA hardware interception and bug-planting workstation*. http://dailycaller.com/2014/05/15/leaked-photos-show-nsa-hardware-interception-and-bug-planting-workstation/.
26. Snyder, B. (2014). *The NSA planted backdoors in cisco products*. http://www.infoworld.com/article/2608141/internet-privacy/snowden/u2013the-nsa-planted-backdoors-in-cisco-products.html.
27. Swierczynski, P., Fyrbiak, M., Koppe, P., Moradi, A., & Paar, C. (2016). Interdiction in practice: hardware Trojan against a high-security USB flash drive. *Journal of Cryptographic Engineering, 7*(3), 1–13.
28. Chari, S., Rao, J. R., & Rohatgi, P. (2002). Template attacks. In *International workshop on cryptographic hardware and embedded systems*. Berlin: Springer.
29. Hospodar, G., Gierlichs, B., De Mulder, E., Verbauwhede, I., & Vandewalle, J. (2011). Machine learning in side-channel analysis: A first study. *Journal of Cryptographic Engineering, 1*(4), 293–302.
30. Polylactic acid (PLA) (2017). http://reprap.org/wiki/PLA.
31. Belikovetsky, S., Yampolskiy, M., et al. (2016). dr0wned-cyber-physical attack with additive manufacturing. arXiv preprint:1609.00133.

Part II
Data-Driven Defense of Cyber-Physical Systems

Chapter 4
Data-Driven Defense Through Leakage Minimization

4.1 Introduction

In Chaps. 2 and 3, data-driven modeling was performed to demonstrate how novel attack models utilizing the cross-domain relationship of cyber-physical system can be used to breach the confidentiality of the system. Moreover, authors in [1–3] have demonstrated how various analog emissions such as *acoustic*, *electromagnetic*, *vibration*, and *thermal* can behave as side-channels and reveal valuable intellectual property (IP) [4] (such as geometry information) during the printing stage. These side-channels allow attackers to acquire the cyber-domain information (such as the printing instructions) without using the brute force approach.

There are various research works that have tackled the issue of *integrity* [5–8], however, at the time of writing this chapter, only a few have worked on *thwarting* the side-channel attacks. In cyber-physical manufacturing systems, attackers are motivated to breach the security of manufacturers for their *Intellectual Property* (IP) and *operations information* that is worth a large amount [9]. Although a large body of work has been focused in preventing duplication of the printed 3D object [10, 11] to maintain *confidentiality*, in this chapter we tackle the problem of IP loss during prototyping stage or early stage of product design, before the 3D object is accessible to an attacker. This stage can be crucial for a company as any information leakage at this stage can cause the company to lose IP to competitors [12].

4.1.1 Motivation for Leakage-Aware Security Tool

In cyber-physical additive manufacturing, critical information that needs to be protected are the intellectual properties hidden in the geometry of the 3D objects, machine-specific processes, etc. [4]. In this chapter, the unique geometric properties

© Springer Nature Switzerland AG 2020
S. R. Chhetri, M. A. Al Faruque, *Data-Driven Modeling of Cyber-Physical Systems Using Side-Channel Analysis*, https://doi.org/10.1007/978-3-030-37962-9_4

of the object, which give their product an edge over other competing products in the market, are considered as the crucial IP. These geometric designs are eventually described and stored in the form of cyber-domain data, which flows through the digital process chain. The digital process chain of additive manufacturing consists of computer-aided design (CAD) tools for modeling the 3D objects, and computer-aided manufacturing (CAM) tools for converting the 3D models to slices of 2D polygons [13], and then generating tool-path (G/M-codes) based on those 2D polygons [14]. These G/M-codes are eventually converted to control signals that actuate the physical components. During actuation, mechanical and electrical energies flow through the system and may leak the information about the G/M-codes, eventually leaking the IP. An attacker may use multi-channel fusion [15] attacks to reverse engineer the 3D geometry of an object.

Information leakage from the side-channels may be prevented by reducing the mutual information between the cyber-domain data and the physical emissions. This may be achieved by changing the mechanical structure of the 3D printer to make emissions corresponding to multiple G/M-code actuation identical, adding noise generators to reduce the signal-to-noise ratio in the individual side-channels, or adding secure chambers. Compared to these measures which add cost, it is possible to change CAM tools to minimize the information leakage from the acoustic side-channel. The CAM tool can change process and design parameters in slicing and tool-path generation algorithm without affecting the quality of the 3D object. Moreover, costly countermeasures for intrusively changing the mechanical structure or noise generating sources are avoided.

4.1.2 Problem and Challenges

Designing a methodology to minimize information leakage in the physical domain through the incorporation of security aware solutions in the cyber-domain of the cyber-physical additive manufacturing system poses the following key challenges:

1. Determining the design variables in cyber-domain (computer-aided manufacturing tools such as slicing and tool-path generation algorithm) that can be optimized to minimize the information leakage.
2. Formulating an optimization problem that can be placed in the digital process chain, which can be generalized for all side-channels, and can balance the trade-off between the design variables and the associated costs (*leakage amount, printing time*, etc.).
3. Understanding the implication of information leakage from the side-channels.

4.1.3 Contributions

To address the above-mentioned challenges, a novel methodology is proposed, which is capable of generating information leakage-aware cyber-physical additive manufacturing tool, and consists of the following:

1. **Leakage modeling and quantification for an additive manufacturing system** (Sect. 4.2.1), which highlights the relationship between the G/M-codes and the emissions in the side-channel, and performs information quantification using *mutual information*.
2. **Formulation of an optimization problem** (Sect. 4.2.3), which describes various design variables (orientation θ and travel feed-rate v) to optimize, and provides it as an input to the slicing algorithm and the tool-path generation algorithm in the digital process chain.
3. **A case study analysis**, for which we formulate the success rate (Sect. 4.2.4), provide an attack model (Sect. 4.2.2), and reconstruct the 3D object (Sect. 4.4.2) to visualize how leakage-aware CAM tool obstructs the reverse engineering of 3D objects from the side-channels.

4.2 System Modeling

The proposed leakage-aware computer-aided manufacturing tool is presented in Fig. 4.1. Traditional digital process chain lacks the feedback from the physical domain, which can convey knowledge about the amount of information leaked from the side-channels. In our work, CAM tools are not only aware of the leakage but also change the design variables to minimize the leakages. The leakage quantification

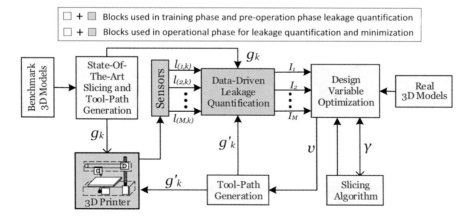

Fig. 4.1 Data-driven leakage-aware computer-aided manufacturing tool

may be done both in design-time and run-time. During design-time, the state-of-the-art slicing and tool-path generation algorithms are used to generate G/M-code $(g_1, g_2, \ldots, g_k)$ corresponding to the benchmark 3D models.

These G/M-codes are sent to the 3D printer and the corresponding analog emissions, such as acoustic, power, vibration, and magnetic, are collected using the sensors. The leakage values $(l_{(M,k)})$ from each side-channels (assuming there are M side-channels) and the corresponding G/M-codes $(g_1, g_2, \ldots, g_k)$ are then passed for leakage quantification. The leakage quantification block calculates mutual information (I_M) corresponding to each of the side-channels. These values are passed to the design variable optimization block, which optimizes the design variables (such as γ, v, etc.) to minimize the mutual information. Then real 3D models (which contains the IP) are passed through the design variable optimization block which modifies the slicing algorithm and tool-path generation algorithm to produce new G/M-codes $(g_1', g_2', \ldots, g_k')$. During run-time, the leakage-aware CAM tool can be constantly updated to quantify the leakage and optimize the design variables. This may be necessary to make sure that the environmental condition and aging of the physical system do not aid in the leakage of the information.

4.2.1 Data-driven Leakage Modeling and Quantification

For modeling the leakage, it is assumed that there are M side-channels. The G/M-code is the *sensitive variable,* that an attacker seeks to extract from the 3D printer. Let G represent the *sensitive* discrete random variable, with probability distribution function $p(g)$, where $g_1, g_2, \ldots, g_k$ represents the possible G-code instructions. Then the leakage from each channel can be written as follows:

$$L_i = \delta_i(G) + N_i \quad i = 1, 2, \ldots, M \qquad (4.1)$$

where N_i denotes an independent noise (independent from the variable G) in the ith channel, $\delta(.)$ represents the deterministic function, and L_i is the leakage in the ith channel. Moreover, for each G-code instruction g_k, the corresponding leakage may be given as follows:

$$l_{(i,k)} = \delta_i(g_k) + n_{(i,k)} \quad k = 1, 2, \ldots, K \qquad (4.2)$$

where $n_{(i,k)}$ represents the leakage noise value in the ith channel for the kth leakage measurement, and K is the total number of G-code instructions. *Mutual information* is used as a *metric* to quantify the information leakage from each of the channels independently. Given that the *joint probability distribution function* $p(g, l_i)$, marginal probability distribution $p(g)$ and $p(l_i)$ for the discrete random variables G and L_i are available or can be estimated, the mutual information between the G-code instruction and the leakage can be calculated as follows:

$$I(G; L_i) = \sum_{l_i \in L_i} \sum_{g \in G} p(g, l_i) \, log_2 \left(\frac{p(g, l_i)}{p(g)p(l_i)} \right) \qquad (4.3)$$

Here, $p(g) = \sum_{l_i} p(g, l_i)$ and $p(l_i) = \sum_g p(g, l_i)$. The joint probability distribution may be calculated empirically $\hat{p}(g, l_i)$ by acquiring the leakage values in the side-channels to various G/M-codes that are produced by the CAM tool for various 3D objects. Moreover, a data-driven modeling approach may be used to estimate the joint probability distribution as the empirical probability distribution tend to overestimate the mutual information. Furthermore, since base two is used for the logarithm, the unit of the mutual information is *bits*.

4.2.2 Attack Model

The side-channel based attack model presented here for a case study is based on earlier works [1, 2, 16] that have successfully demonstrated the possibility of side-channel attacks on 3D printers. In the adversary/attack model, the main objective of an adversary is to be able to steal/infer the intellectual property inherent in the geometry of the 3D objects, which is described by the G/M-codes. As a medium, they will utilize the emissions from the 3D printer, which may behave as a side-channel. The tools used by the attackers in analyzing the side-channels and reverse engineering the G/M-codes are machine learning algorithms which allow them to estimate leakage model function from the emissions. An attacker may be a weak or a strong attacker, depending on their domain knowledge (layer wise printing, a high correlation between two consecutive layers, a dimension of the 3D printer, machine-specific M-codes, etc.). This demarcation is necessary, as geometry information may not be fully stolen from the side-channel. A weak attacker will rely on just the side-channels to estimate the G/M-codes. However, a strong attacker may be able to post-process the estimated G/M-code using domain knowledge to improve the accuracy of the reconstructed 3D objects. For strong attackers, a partial reconstruction of the geometry from itself may be enough to infer further geometry details, hence any information from the side-channel may help them in stealing partial or complete intellectual property.

Furthermore, an attacker goes through two stages to steal the cyber-domain information (G/M-code) (see Fig. 4.2). The first is the training stage, in this stage, we assume that an attacker has high-level access to a 3D printer with a similar physical structure to the target 3D printer. By high-level access, we mean that he/she is able to access the digital process chain (CAD and CAM tools). Hence, the attacker is able to collect the analog emissions from M side-channels corresponding to the G/M-codes. Then the attacker uses data-driven modeling (using machine learning algorithms) approach to estimate the leakage model function $\hat{f}(., \alpha)$, which is used to estimate the relation between the leakage and the G/M-code, where α is the tuning parameter for data-driven models. Moreover, the attacker can extract various features and

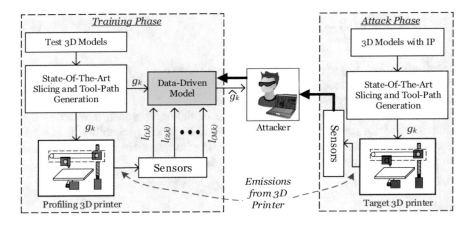

Fig. 4.2 Side-channel attack model for 3D printer

perform feature selections from multiple side-channels to estimate the leakage
model function. Due to the presence of a large number of correlated features from
multiple side-channel, the attacker is assumed to perform feature reduction (using
methods such as principal component analysis, singular value decomposition, etc.).
In doing so, they perform automatic selection of relevant features from multiple
side-channels for reverse engineering the G/M-code. Using this function, an attacker
can estimate the G/M-code based on the analog emissions from the side-channel.
$\hat{g}_k = \hat{f}(l_{(1,k)}, l_{(2,k)}, \ldots, l_{(M,k)})$. Various machine learning algorithms can be used
to model the *leakage model function* $\hat{f}_i(., \alpha)$, such that

$$i = \arg\min_{1 \leq i \leq N} \sum_{k=1}^{K} |g_k - \hat{f}_i(., \alpha)| \tag{4.4}$$

where N is the various *leakage model function* estimated by the attacker. The
accuracy of the estimated function depends on the amount of information leaked
about G/M-code in the side-channels. We assume that an attacker has unlimited
access to the profiling 3D printer during the training phase and can acquire a large
number of leakage values corresponding to G/M-codes to build the leakage model
functions. The next phase is the attack phase. In this phase, he/she does not have
high-level access to the digital process chain of the 3D printer. However, they are
able to surreptitiously place sensors to monitor analog emissions with low-level
physical access. By low-level access, we mean that they are not able to have digital
access to the process chain, however, may be able to have closer non-intrusive
physical access to the 3D printer. A malicious insider with low-level access may
be an ideal point of contact for an attacker to place the sensors and acquire the
analog emissions. Using these emissions, an attacker utilizes the estimated leakage
model functions to predict the G/M-codes.

4.2.3 Formulation of Data-Driven Leakage-Aware Optimization Problem

In this section, we will formulate the leakage-aware optimization problem for CAM tools used specifically for the fused deposition modeling technique based 3D printers, and show how we can provide feedback to the CAM tool for optimizing various design variables for minimizing the information leakage.

4.2.3.1 Design Variables for Leakage Minimization

The fundamental principle behind reducing the leakage from the side-channels includes a selection of the design variables which can be optimized without substantially affecting the geometry of the 3D objects being printed. This variable may be selected from various stages, either the CAD tools or CAM tools. Since the optimization algorithm is incorporated in the CAM tool, design variables are selected from the slicing algorithm and tool-path generation algorithm for minimizing the leakage. The exhaustive list of design variables is not presented here, however, the proposed methodology can be scaled when other design variables are added. Here, two design variables (γ, v) are proposed.

Assumption 1 *For the 3D printer, while printing in each layer (the XY-plane), let v be the velocity of the nozzle head (or the extruder from where the filament is deposited) and θ be the angle made by the line segment being printed with the X axis. Let $g_i(\theta)$, be a function that gives the mutual information between the G/M-code (with varying v) and the analog emissions from the ith side-channel when the angle between the line segment and the X axis is θ. Then for each side-channel there exist angle β_i such that $\beta_i = \arg\min_\theta g_i(\theta)$, where $0 \leq \beta_i \leq 2\pi$, and $i = 1, 2, \ldots, M$.*

Remark 1 For each of the side-channels, changing the angle of the line segment θ has the corresponding effect of increasing the source of emission. For example, when the angle is $\theta = 0°$ corresponds to line segment being printed parallel to X axis. Hence, a minimum of X and extruder's stepper motor have to be active. When $\theta = 45°$, line segment is printed with equal speed v in both X and Y axes. Hence, a minimum of three stepper motors has to be active (X, Y, and the extruder). Due to the varying number of source of emission added with varying θ, mutual information for different θ will vary accordingly. Moreover, mutual information will be low if the complex axis movements leakages cannot be distinguished. Hence, there may exist a certain angle for which minimum $I(G; L_i)$ value can be obtained.

In 3D printing, there is a large number of straight line segments, optimizing the θ for each segment to minimize leakage can affect the convergence of the optimizing algorithm. Rather, principal component analysis (PCA) is used to find the common orientation angle of all the line segments. In the digital process chain, the 3D model is converted to a file with tessellated triangles that describe the geometry. From this file, using the cross product, a vector, u, normal to the plane of the triangular surface is calculated. Next, PCA is performed on the collection of vectors u calculated for all the triangular surfaces. Then, the first principal component which has the highest eigenvalue is extracted. This value represents the most common normal vector of all the line segments. Here, γ is defined as the angle of the vector u' which is perpendicular to the first principal component of the vector u. This is the first design variable.

Assumption 2 *Given the nozzle movement to print a line segment of length l in xy-plane, let v_x and v_y be its velocity in x and y axes, respectively, where travel feed-rate is $v = \sqrt{v_x^2 + v_y^2}$. Then there exists $v_i' \in \mathbb{R}$ such that $v_i' = \arg\min_v g_i(v)$, where $g_i(v)$ gives the mutual information between the G/M-code (with varying θ) with speed v and the analog emissions from the ith side-channel.*

Remark 2 v' values that will achieve the minimum mutual information in the side-channel lie in the higher travel feed ranges. Considering the acoustic side-channels, first the higher frequency excitation due to faster travel feed-rates will cause a reduction in the amplitude of the vibration as most of the time this excitation force acts in opposing the direction of the vibration, and second, the leakage signal will be corrupted quickly by new analog emission from the next G-code. Due to this, the sample of data collected for the G-codes with large travel feed-rate will be less in number and may be contaminated by another G-code leakage signal. Hence, due to the mixture of the leakage signals for different G-code, the mutual information extracted will be low.

State-of-the-Art CAM Tools Current slicing algorithms for fused deposition modeling based desktop 3D printers do not consider the information leakage through the side-channels and have tool-path generation that is optimized only for machining efficiency (such as time, material deposition, precision, etc.).

4.2.3.2   Optimization Problem Statement

For minimizing the information leakage from side-channels, based on assumptions 1 and 2, a new leakage-aware algorithm is proposed. The design variables are defined as, $0 \leq \gamma \leq 2\pi$, and $v = \sqrt{v_x^2 + v_y^2}$, where $v_x \in \mathbb{R}$, and $v_y \in \mathbb{R}$. For the speed in x and y axes, the two variable bounds are $v_{xmin} \leq v_x \leq v_{xmax}$ and $v_{ymin} \leq v_y \leq v_{ymax}$, where v_{xmin} and v_{ymin} are the minimum machine-specific travel feed-rate in x and y axes, respectively, and v_{xmax} and v_{ymax} the maximum machine-specific travel feed-rate in x and y axes, respectively. Here, a simple constraint $T \leq kT_{orginal}$ is used, where $T_{original}$ is the printing time of the state-of-the-art slicing and tool-path generation algorithm, and $k \geq 1$ is the user defined constant. The mutual information between the G-code and the leakage signal $I_{\gamma_i}(G; L_i)$ and $I_{v_i}(G; L_i)$ is calculated either at design-time or at run-time by collecting the various analog emissions corresponding to the G/M-codes of various 3D objects. Then, using a non-linear polynomial functions $f_{\gamma_i}(I_{\gamma_i}, \gamma_i)$ and $f_{v_i}(I_{v_i}, v_i)$, the relation between the mutual information and the design variables in different side-channels (acoustic, vibration, power, and electromagnetic) can be estimated. Then, for reducing the mutual information between the analog emission and the design variables, the multi-objective optimization function can be given as follows:

$$(\gamma, v) = \underset{(\gamma, v)}{\arg \min} \left(f_{\gamma_1}, f_{\gamma_2}, \ldots, f_{\gamma_M}, f_{v_1}, f_{v_2}, \ldots, f_{v_M} \right) \qquad (4.5)$$

Based on the value given by the optimized design variable, slicing and tool-path generation will generate new G-code with minimum information leakage. The algorithm to generate G/M-codes is shown in Algorithm 1. The input to the algorithm is the probability distribution $(\hat{p}_i(\gamma, l_i), \hat{p}_i(v, l_i))$ estimated by collecting the leakage and the G/M-code data while printing the benchmark 3D models, and STL file of the original 3D objects. In line 1, step size for estimating the cost function based on the design variables γ, v is defined, along with their range. Then from line 2–6, using the probability distribution $\hat{p}_i(\gamma, l_i), \hat{p}_i(v, l_i)$, various mutual information values are calculated for the varying design variables.

Algorithm 1: Data-driven leakage-aware G-code generation

  Input: Empirical or data-driven joint distribution $\hat{p}_i(\gamma, l_i), \hat{p}_i(v, l_i)$
  Input: Output of CAD Tool (`STL Files`)
  Output: Modified G-code g'
1 Define step size $\triangle_\gamma, \triangle_v$ and range min_γ, min_v and max_γ, max_v
2 **for** $i = 1 : M$ **do**
3   **for** *each* $j \in (\gamma, v)$ **do**
4   **for** $k = min_j : \triangle_j : max_j$ **do**
5             $I_{(j_i, k)} = I_k(G; L_i)$ // Based on $\hat{p}_i(\gamma, l_i)$
                                                                          // and $\hat{p}_i(v, l_i)$
6   Estimate Non-linear function f_{j_i}
7 Optimize $\arg \min_{(\gamma, v)}(f_{\gamma_1}, f_{\gamma_2}, \ldots, f_{\gamma_M}, f_{v_1}, f_{v_2}, \ldots, f_{v_M})$
8 $g' =$ SliceandToolPathGeneration$(\gamma, v,$ `STL File`) **return** g'

In line 6, polynomial function is used to estimate the relation between the design variables and the mutual information calculated in line 5. Then based on the description of the problems statement, mixed multi-objective non-linear integer programming is used to optimize the design variables. In line 8, the modified design variables are passed to the slicing and tool-path generation function to generate a G-code with minimum leakage, which is finally returned in line 9.

4.2.4   Success Rate of the Adversary

Mutual information provides an idea about how much information is available to an attacker for the exploitation [17]. However, it does not provide information about the capability of an attacker. A strong attacker may be able to exploit little information for a successful attack, whereas a weak attacker may not be able to exploit the information available for the attack. Hence, in addition to the mutual information, the success rate of an attacker helps in demonstrating how side-channels can be leveraged for breaching the *confidentiality* of the 3D printer. Hence, compared to the work in [18], in this chapter, the success rate based on the adversary model presented in Sect. 4.2.1 is also analyzed. It is assumed that an attacker has unlimited access to a replica of the manufacturing system, and by using state-of-the-art learning algorithms he/she is able to generate data-driven leakage model of the system.

3D printer instructions can be divided into two types [19]. The first type is called G-codes and they are responsible for the movement of the nozzle in different axis. The second type is called M-codes, and they are responsible for controlling the parameters of the machine beside motion. An example of a G type instruction is a command which results in moving the nozzle of a 3D printer for a certain distance, with a certain angle and speed. On the other hand, an example of the M type instructions is an instruction which sets the acceleration rate for the stepper motor. Here, for the sake of simplicity, it is assumed that the number of M type instructions are limited, and do not impact the geometry of the 3D objects [19], which is the case in 3D printers. Hence, the focus is given only to G type instructions for measuring the success rate. The G type instructions or the G-codes consist of instruction such as (1) *movement axis*, (2) *distance moved in each axis*, (3) *speed of motion in each axis*, and (4) *extrusion amount*. The extrusion amount is calibrated to maintain the optimal print quality and is fixed in most of the CAM tools. It does not affect the geometry of the object being printed. The Z axis movement occurs every layer with constant height and can be distinguished or inferred with ease. In state-of-the-art slicer used in the CAM tools, such as Cura [20], speed does not change dynamically. Rather, most of the time, it has constant predefined values for each stage (*first layer, outer boundary, inner filling, and traveling without printing*) of printing. Since speed variation in each of the stages is drastically different, they are easier to classify. Hence, the most challenging information that an attacker may require to extract from a G-code is the geometry of the object being printed in each XY-plane. This information can be abstracted in terms of angle and length of the line segment being printed by each G-code.

Based on these assumptions, the G-code instruction is reduced to $E_{(\theta,d)}$, where $0 < \theta < 360$ is the angle between the movement direction of a line segment and the X axis, d is the travel length. Then for calculating the success rate of an attacker, only measure of how successfully he or she is able to reconstruct θ and d for each line segment is required. Let us represent the sequence of G-codes as $\{E_1, E_2, \ldots, E_i\}$, with $E_i = E_{(\theta_i, d_i)}$. Let L_i be the leakage corresponding to E_i.

Then an attacker will utilize the function $\hat{f}(., \alpha)$ estimated using Eq. 4.4 to infer about E_i based on L_i. Estimated $\hat{E}_i$ consists of $\hat{\theta}$ and $\hat{d}$.

In our attack model, attacker estimates $\hat{\theta}, \hat{d}$ independently. Hence, there are two functions $\hat{\theta} = \hat{f}_\theta(., \alpha)$ and $\hat{d} = \hat{f}_d(., \alpha)$. Now, the success rate is formulated in terms of accuracy in reconstruction of each of the line segment using $\hat{\theta}$ and $\hat{d}$.

For doing so, let (x_1, y_1) and (x_2, y_2) be a line segment a G-code instruction prints on a XY-**plane**. As explained earlier, this segment can be explained in terms of polar coordinates using θ and d. Now, let the estimated polar coordinate value be $\hat{\theta}$ and $\hat{d}$ as shown in Fig. 4.3. Then, the success rate is defined in terms of a parameter l, which can be given as follows:

$$l = \sqrt{d^2 - 2dd'cos(\delta_\theta) + \hat{d}^2} \tag{4.6}$$

Then the success rate is measure of how accurate is l, given as follows:

$$Success\ Rate\ (SR) = \frac{1}{n} \sum_{i=1}^{n} S_l \tag{4.7}$$

where $S_l = 1$ when $|\hat{l}_i - l_i| < e_l$ and 0 otherwise, and n is the total number of G-code instructions being reconstructed. To prove that success rate encompasses accuracy estimation of both θ and d, three cases are considered. First, let angle estimation be 100% accurate. In that scenario, $\delta_\theta = 0$, then $l = \pm(d - \hat{d})$ from Eq. 4.6, which shows that l still depends on the estimation of d. In second case let us assume that estimation of d is 100% accurate, then $d = \hat{d}$. Then Eq. 4.6 reduces to $l = \pm d\sqrt{2 - 2cos(\delta_\theta)}$. It shows that in such scenario l still depends on estimated value of θ, as $\delta_\theta = \hat{\theta} - \theta$. In the final case, let estimation of θ and d be 100% accurate, then $\delta_\theta = 0$ and $\hat{d} = d$. Then from Eq. 4.6, $l = \sqrt{d^2 - 2ddcos(0) + d^2} = \sqrt{d^2 - 2d^2 + d^2} = 0$, making success rate 100% accurate as well. Hence, this metric will give the total success rate of the G-code instruction in terms of $\hat{\theta}$ and $\hat{d}$.

Fig. 4.3 Success rate formulation for each G-code instruction

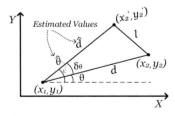

4.3   Experimental Results

The experimental setup for the leakage-aware computer-aided manufacturing tool is shown in Fig. 4.4. The setup consists of a fused deposition modeling based desktop 3D printer [21]. Three AT2021 cardioid condenser audio sensors [22] are placed parallel to X, Y, and Z axes, respectively, and treated as individual channels to consider sound emanation in multiple direction. The vibration sensor (Adafruit triple-axis accelerometer [23]) is placed at the static part of the 3D printer, measuring vibration in all three axes. A current clamp [24] is used to non-intrusively measure the current being passed to the main circuit board of the printer from a power supply. A Honeywell's magnetometer (HMC5883L [25]) is used to measure the magnetic fluctuation around the 3D printer. Beside the current clamp, most of other sensors (audio, vibration, magnetic, etc.) are all present in modern smartphones, and have been used in the attack model presented in [1] for breaching the *confidentiality*.

Since four side-channels are used, here, $M = 4$. Audio sensors placed in different axis are fused together and treated as a single side-channel. From the raw signals collected from these sensors, power spectral density in the frequency range of 60 Hz–10 kHz is calculated, principal components of the corresponding power spectral density are analyzed, and the mutual information between these principal components and the design variables are measured. In our experiment, we have performed leakage quantification and mutual information calculation only during design time. This means, the mutual information for various design variable are not updated at run-time.

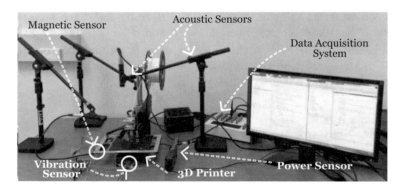

Fig. 4.4   Experimental setup with multiple sensors

4.3.1 Mutual Information

In this section, analysis between the mutual information and the design variables is presented. This will demonstrate how the non-linear function in line 6 of Algorithm 1 can be estimated.

4.3.1.1 Design Variable—γ

γ is varied from 0 to 90° with the step size $\triangle_\gamma = 10°$. For estimating the joint probability function $\hat{p}(\gamma, l_i)$, γ is distributed uniformly, and the corresponding analog emissions from the side-channels are collected. This joint probability function $\hat{p}(\gamma, l_i)$ is then used in calculating the mutual information using Eq. 4.3.In Fig. 4.5, mutual information between the first principle component of the power spectral density values of the four analog emissions and the design variable γ are presented. The total entropy of the design variable (γ) is $log_2(11) \cong 3.4594$ bits. It can be observed that for the design variable γ, vibration leaks more information compared to the other emissions. For motion in XY-plane, since v is constant for printing a single line segment, varying γ changes the v_x and v_y instead (as $v = \sqrt{v_x^2 + v_y^2}$).

This means varying γ results in actuation of stepper motor moving in X and Y axes with the different angular speed. For vibration side-channel, change of angular speed of stepper motors will result in a change of acceleration measured in X and Y axes, immediately. The acoustic side-channel would acquire this variation in the acceleration of sound pressure only when the surface of the 3D printer has been excited. This causes the vibration side-channel to pick more information about γ variation than acoustic side-channel. The low-field magnetic sensors used in the experiment measures how the mechanical components of the 3D printer affect the

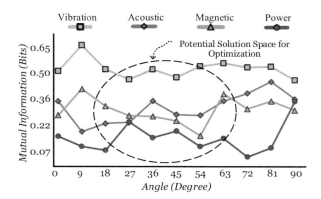

Fig. 4.5 Mutual information between angle (γ) and leakage

strength and directional of the earth's magnetic fields. Variation in γ causes the metallic part of the FDM 3D printer to cut the earth's magnetic field at a varying rate. Hence, the variation of γ may be inferred from the magnetic side-channel. The power side-channel has lowest mutual information with a variation of γ. The power signal is one-dimensional compared to another emission which measure the signal variation in more than one dimension (X, Y, or Z axis). Hence, it is intuitive that it leaks less information compared to the other side-channels. From Fig. 4.5, it may also be observed that single axis movement (when $\gamma = 0$ and $\gamma = 90$) has more mutual information and thus leak more information compared to multiple axis movements (as there is emission from multiple components of the 3D printer).

4.3.1.2 Design Variable—v

The travel feed-rate (v) is varied from 700 to 3300 mm/min (which is the range for the given 3D printer) with the step size $\triangle_v = 200$ mm/min, to demonstrate the relation between the design variable v and the mutual information. Similar to γ, the joint probability function $\hat{p}(v, l_i)$ is estimated by collecting the analog emissions for the uniformly distributed travel feed-rate values.

The total entropy of the design variable (v) is $log_2(14) \cong 3.8074$, as there are 14 speed values. In Fig. 4.6, the mutual information between the first principal component (with the highest eigenvalue) of the power spectral density and the varying travel feed-rate for vibration, acoustic, power, and magnetic emissions is presented. When the travel feed-rate is increased in the G/M-code, the stepper motor driver increases the rate of current passing through the electromagnetic stator core of the stepper motor. The presence of a permanent rotor and the electromagnetic stator with a varying rate of constant current supply makes stepper motor capable of producing audible sound. Moreover, the frame in contact with the stepper motor helps the vibration to be conducted throughout the mechanical frame of the 3D printer, and allows the vibration side-channel to pick subtle variation even before the sound is emanated in the air. Moreover, the varying rate of current supply fluctuates the power consumption in each of the stepper motors used for actuating

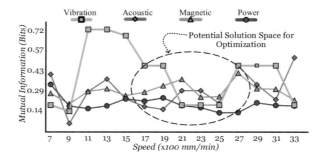

Fig. 4.6 Mutual information between speed (v) and leakage

the G/M-code. The changing current rate affects the magnetic field created in the stator, however, the sensor used in our experiment is not able to capture these small variations of the magnetic field. However, the varying travel feed-rate changes the rate of movement of the mechanical structure, which affects the magnetic field of the earth. This can be captured by the magnetic sensors. From Fig. 4.6, it can be observed that, for speed variation, vibration side-channel reveals more information compared to other side-channels. As explained earlier, the vibration side-channel is able to capture the minute variation in vibration caused by the variation of the travel feed-rate before the acoustic side-channel, resulting in it having higher mutual information than the acoustic side-channel. Moreover, the mutual information is higher for the lower speeds, lower for the medium range speeds, and again higher for the faster speeds. This may be due to the fact that the 3D printer has certain resonant frequencies caused by the cumulative movement and vibration of the individual components, and the speed ranges causing the vibration closer to resonant or harmonics of the resonant frequencies leak more information. The power side-channel has lower mutual information compared to others, this may be due to the fact that the base plate heater and the heater in the nozzle of the 3D printer consume larger current, and variation of stepper motor current fluctuation has lower SNR causing lower mutual information between the power leakage values and the G/M-code.

The optimization algorithm utilizes the non-linear relation estimated in Figs. 4.5 and 4.6 to minimize the mutual information between the analog emissions and the cyber-domain data to effectively reduce the information leakage.

4.3.2 Test with Benchmark 3D Models

The CAD models of the benchmark models are selected from [26] that cover wide-variation in G/M-codes. The selected benchmark 3D models are shown in Fig. 4.7. Some of these objects (numbered *1, 2, 3, 4, 11*, and *8*) are selected because they are used for calibrating and tuning the 3D printer, whereas others (numbered *5, 6, 7, 9, 10, 12*, and *13*) are selected due to the variation in complexity of the design. For each of these benchmark models, mutual information between the G-codes and each of the side-channels is calculated before and after changing the design variables (γ, v). Figures 4.8, 4.9, 4.10, and 4.11 show the results of mutual information calculation for acoustic, power, magnetic, and vibration side-channels, respectively.

For acoustic side-channel, the average drop of mutual information by **24.94%** is obtained across all the benchmark models (see Fig. 4.8).

However, for object number 10 and 12 (see Fig. 4.7) there is only a small change in mutual information rather than large reduction when the leakage-aware slicing and tool-path generation algorithm is used. This may be due to the fact that objects 10 and 12 consist of geometric structures (spherical and circular) that do not benefit from design variable γ (as the change in γ does not introduce variation in the orientation of the spherical and the circular objects). However, real-world objects

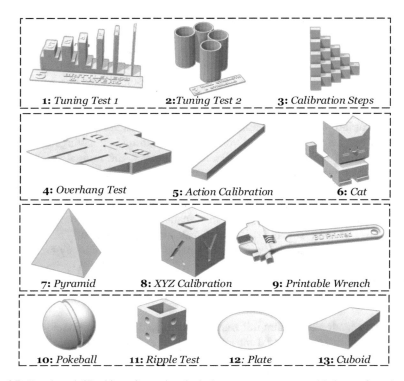

Fig. 4.7 Benchmark 3D objects for testing the leakage-aware computer-aided manufacturing tool

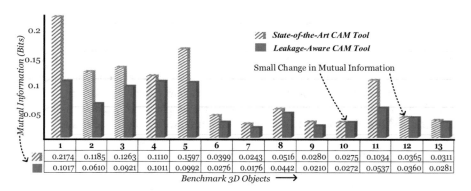

	1	**2**	**3**	**4**	**5**	**6**	**7**	**8**	**9**	**10**	**11**	**12**	**13**
State-of-the-Art	0.2174	0.1185	0.1263	0.1110	0.1597	0.0399	0.0243	0.0516	0.0280	0.0275	0.1034	0.0365	0.0311
Leakage-Aware	0.1017	0.0610	0.0921	0.1011	0.0992	0.0276	0.0176	0.0442	0.0210	0.0272	0.0537	0.0360	0.0281

Benchmark 3D Objects ⟶

Fig. 4.8 Mutual information between G-codes and acoustic side-channel

are not just spherical or circular, and the mutual information will decrease when leakage-aware CAM tool is used.

For power side-channel, the average drop of **32.91%** in mutual information is obtained across all the benchmark models. As shown in Fig. 4.9, the mutual

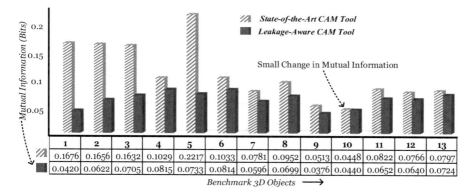

Fig. 4.9 Mutual information between G-codes and power side-channel

	1	2	3	4	5	6	7	8	9	10	11	12	13
▨	0.1676	0.1656	0.1632	0.1029	0.2217	0.1033	0.0781	0.0952	0.0513	0.0448	0.0822	0.0766	0.0797
■	0.0420	0.0622	0.0705	0.0815	0.0733	0.0814	0.0596	0.0699	0.0376	0.0440	0.0652	0.0640	0.0724

Benchmark 3D Objects ⟶

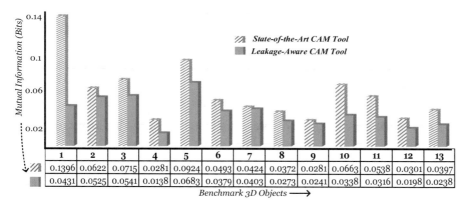

Fig. 4.10 Mutual information between G-codes and magnetic side-channel

	1	2	3	4	5	6	7	8	9	10	11	12	13
▨	0.1396	0.0622	0.0715	0.0281	0.0924	0.0493	0.0424	0.0372	0.0281	0.0663	0.0538	0.0301	0.0397
■	0.0431	0.0525	0.0541	0.0138	0.0683	0.0379	0.0403	0.0273	0.0241	0.0338	0.0316	0.0198	0.0238

Benchmark 3D Objects ⟶

information dropped for all the benchmark models except object 10 (only a small change was observed). Unlike other side-channels, power side-channel measures data only in one-dimensional (unlike acoustic, vibration, and magnetic that measure fluctuation in all X, Y, and Z axes). The variation in the instantaneous power of the 3D printer occurs due to specific activation sequence of various components. For spherical geometry like object 10, due to symmetric property, the variation in γ does not result in a change in the sequence of activation drastically.

For magnetic side-channel, the average drop of **32.29%** in mutual information is obtained across all the benchmark models (see Fig. 4.10). For all side-channels, the objects that benefited the most in reducing the mutual information were object 1 and object 5 (see Fig. 4.7). This is due to the fact that these objects are stretched in either X or Y axis and hence changing the γ drastically changes its orientation, thus affecting the analog emissions.

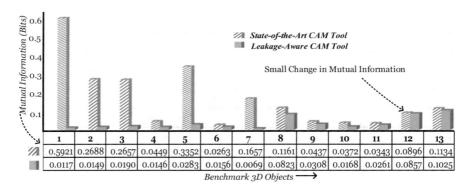

Fig. 4.11 Mutual information between G-codes and vibration side-channel

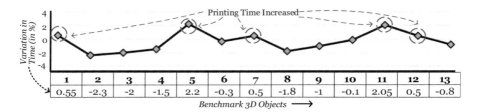

Fig. 4.12 Variation in printing time of leakage-aware CAM tool compared to the state-of-the-art

For vibration side-channel, the average drop of **55.65%** in mutual information is obtained across all the benchmark models (see Fig. 4.11). This drop in mutual information is highest among all the side-channels. This result corroborates the analysis presented in Sect. 4.3.1. Optimizing the design variables γ and v affects the vibration side-channel the most. Provided the fact, there are more design variables and machine-specific process parameters that may be tuned to reduce the mutual information, design space exploration of such variables may be performed to drastically reduce the mutual information from all the side-channels. Similar to acoustic side-channel, it also has a relatively small change in mutual information for object 12.

Figure 4.12 shows the timing variation between the state-of-the-art CAM tool and the leakage-aware CAM tool expressed in terms of percentage increase or decrease in the printing time. It can be seen that for the given benchmark 3D objects, the maximum increase in the printing time was **2.2%**.

4.4   Case Study with an Attack Model

The mutual information provides the system designers about the amount of information that be deduced about the G/M-code from the analog emissions. However, the success rate in reconstructing the G/M-code also highly depends on the capability of an attacker. A strong attacker (with domain knowledge, large computing capability, and some high-level access to the digital process chain of the target 3D printer) may be able to achieve higher success rate with low mutual information. In this chapter, we do not provide an exhaustive analysis of the success rate based on the various capability of an attacker. Rather, we choose the attack model presented in Sect. 4.2.2, which has also been used in previous works [1, 2].

4.4.1   Success Rate Calculation

In the attack model, an attacker will utilize leakage from all the size-channels to infer about various G-codes using a function $\hat{f}(., \alpha)$. Moreover, each parameter of the G-code such as θ and v is estimated using functions $\hat{\theta}_k = \hat{f}_\theta(l_{(1,k)}, l_{(2,k)}, l_{(3,k)}, l_{(4,k)})$ and $\hat{d}_k = \hat{f}_d(l_{(1,k)}, l_{(2,k)}, l_{(3,k)}, l_{(4,k)})$. Since G/M-code consists of various information about the line segment, it becomes intuitive to have multiple functions estimating these parameters, and later combining them together to reconstruct the line segment. As a test case study, a machine learning algorithm called *Random Forest* (previously used in [2]) is used to estimate the functions $\hat{f}_\theta(., \alpha)$ and $\hat{f}_d(., \alpha)$, and the success rate is calculated using Eq. 4.7 for various values of e_l ranging from 0 to 10 mm with step size of 0.5 mm. As mentioned in Sect. 4.2.4, Eq. 4.7, measures the average success rate in reconstructing the line segment defined by the G/M-code. The error threshold e_l corresponds to the combined error in predicting the length d of the line segment and the angle θ made with the X axis. For feature, an attacker extracts the power spectral density values from all the side-channels. These features are then fused together and passed to the random forest algorithms for estimating the functions. By combining the features from all the side-channels, the estimated function effectively selects the channel that gives more information about the d and θ. During the training phase, 75% of the benchmark models are used to gather analog emissions from the side-channel, estimate the functions, and the rest of the benchmark model's G-code is predicted using these functions. Since an attacker has high-level access to the 3D printer, he/she may be able to acquire unlimited training data. However, for feasibility and variation in training data, we have selected the benchmark models and divided it into test and training to mimic the two phases of the attack model.

The average success rate for the reconstruction of the line segments described by the G/M-code for varying ranges of e_l is shown in Fig. 4.13. It can be observed that compared to the state-of-the-art CAM tool the success rate of an attacker in reconstructing the line segment is less in the leakage-aware CAM tool. The success

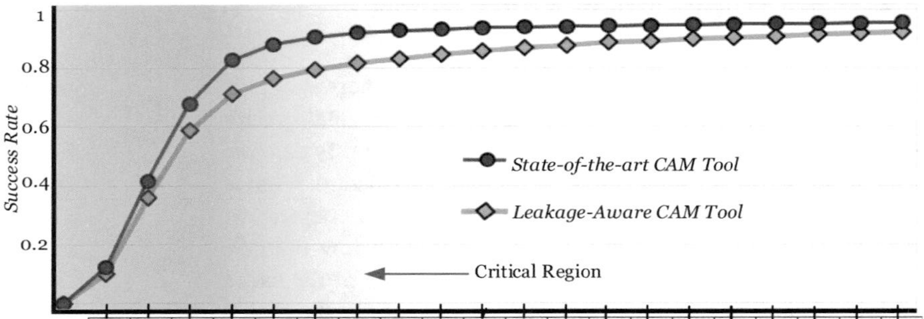

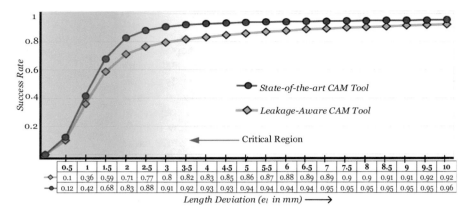

	0.5	1	1.5	2	2.5	3	3.5	4	4.5	5	5.5	6	6.5	7	7.5	8	8.5	9	9.5	10
◇	0.1	0.36	0.59	0.71	0.77	0.8	0.82	0.83	0.85	0.86	0.87	0.88	0.89	0.89	0.9	0.9	0.91	0.91	0.92	0.92
●	0.12	0.42	0.68	0.83	0.88	0.91	0.92	0.93	0.93	0.94	0.94	0.94	0.94	0.95	0.95	0.95	0.95	0.95	0.95	0.96

Length Deviation (e_l in mm) ⟶

Fig. 4.13 Average success rate for G-code reconstruction with varying e_l

rate reduction across the ranges of $e_l = 0$ to 10 mm is 8.74%. Moreover, the reduction in success rate is more than 10%, when $0\,\text{mm} < e_l \leq 5\,\text{mm}$. This is the critical region for e_l, as an attacker may acquire accurate geometry information easily without further post-processing in this range. The success rate calculated here is in reconstructing the individual line segments. This still does not explain the accuracy in overall 3D object reconstruction. For reconstructing the 3D object, an attacker will have to use post-processing to reconstruct the 3D object layer-by-layer, using the estimated line segments. This post-processing depends on the capability of an attacker and is out of the scope of this chapter. However, we argue that with a higher success rate in each of the line segments, an attacker will have to use less post-processing in reconstructing the 3D object, hence reduction of the success rate for each of the line segments still helps in obfuscating the intellectual property of the object.

4.4.2 Test Case with Reconstruction

In this section, simple post-processing is applied to visualize the effect of a decrease in mutual information and success rate on the reconstruction of the 3D objects (after the G-code prediction). The predicted line segments described by G/M-code from the previous section are used for the 3D object reconstruction. This procedure is carried out for the G/M-codes produced by the state-of-the-art CAM tool (Cura), and the leakage-aware CAM tool. The two test objects chosen have relatively simple (for action calibration or bar) and complex (for a printable wrench) structures. The result of the reconstruction of the test objects for leakage-aware and leakage-unaware CAM tool is shown in Fig. 4.14. The post-processing used involved using the same M-codes generated for the training objects on the test objects as well. During a real

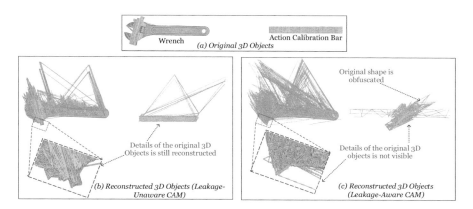

Fig. 4.14   Reconstructed and traced G-codes of test case objects

attack, an attacker is assumed to have some information about the M-codes of the target 3D printer. As M-codes are machine-specific parameters such as extruder and bed temperature, alignment of the printing bed, etc. These parameters are normally set to an optimal constant value for a given 3D printer. This makes sure the quality is maintained. And attacker, during the training phase, has access to the digital process chain of the profiling 3D printer, which is assumed to be similar in build to the target 3D printer. Hence, he/she can use the machine-specific information as a post-processing step for reconstruction of the 3D objects.

From Fig. 4.14, it can be observed that the objects reconstructed from emission captured from the side-channels, when state-of-the-art CAM tool is used, consist of a relatively higher amount of details. On the other hand, the geometry details of the objects are not reconstructed when the leakage-aware CAM tool is used. Without using the leakage-aware CAM tool, it can be observed that in *Printable Wrench*, the details of the outline can still be reconstructed. The obfuscation correlates to the amount of decrease in the mutual information and the success rate. For instance, the action calibration bar (object 5), which has a larger drop in mutual information (compared to object 9), caused the reconstruction to be distorted (making it look nothing like the original shape). This can limit the attacker's capability to steal the valuable intellectual property inherent in the detailed geometry design, thus preventing intellectual property theft.

4.5   Discussion

Positioning of the Sensors In this chapter, the sensors position, number, and orientation have not been explored. An effective attacker may surreptitiously place multiple sensors for data acquisition which may aid in reconstruction. Moreover, to explore the position of sensors to acquire better emission signals may aid in improving the performance of leakage-aware CAM tool as well. Currently, the sensors have been placed in a location that is non-intrusive to the printing process, such as the stationary base of the 3D printer for vibration sensor, stable ground for magnetic sensors, and X, Y, and Z axes for acoustic sensors.

3D Printer Variation Since the attack model relies on analog emissions that behave as side-channels, it highly depends on the mechanical structure of the 3D printer. In literature, we have found that the researcher has tested the effectiveness of the attack model using fused deposition modeling based 3D printers, especially the system with Cartesian movement. Since these printers have similar mechanical build, the same approach for estimating the G/M-code may work; however, this needs to be tested for 3D printers with other mechanical build types (such as polar, delta, etc.) and technologies (such as stereolithography, selective laser sintering, digital light processing, etc.)

Optimization Constraints and Design Variables The design variables such as orientation γ and speed v do not increase the printing time of the 3D model. However, the test case reconstruction shows that some geometry information is still reconstructed even after the leakage-aware CAM tool is used. Hence, more design variables need to be explored in the additive manufacturing domain. These design variables, however, cannot increase various optimization constraints or affect the quality of the 3D objects. Moreover, the same design variables may be able to decrease the mutual information drastically provided the optimization constraints are relaxed. In this way, the trade-off between *confidentiality* and the optimization constraints (such as printing time) may be handled. Furthermore, we demonstrated the possibility of a reduction in mutual information and the success rate in the reconstruction of the 3D object using just two design variables. We acknowledge, that 10% reduction in success rate may not be able to completely hide the geometry (specifically the larger outlines in Fig. 4.14) of the 3D objects. However, provided the fact that there are many design variables (both in object specific G-codes and machine-specific M-codes), our methodology can easily be extended to achieve a higher level of obfuscation in geometry.

Formulation of Success Rate for IP Theft We presented a success rate based on the reconstruction of the individual G/M-code line segments. However, for the success rate in stealing the intellectual property of the 3D object, a proper metric incorporating the capability of an attacker is necessary. Nonetheless, the proposed leakage-aware CAM tool helps in reducing the mutual information and increasing the cost of reconstruction for an attacker.

Other Means of Reduction of Information Leakage This chapter lacks a comparison of the results with other solutions. However, to the best of knowledge this is the first approach in reducing the information leakage from the side-channels in a cyber-physical additive manufacturing system, without additional system modification, and thus no additional cost. However, intrusive system modification (like the addition of noise generators) may achieve better results with added cost.

4.6 Summary

In this chapter, we presented a novel data-driven methodology to improve the *confidentiality* of the cyber-physical manufacturing systems. We achieved this by incorporating leakage-aware computer-aided manufacturing tools such as *slicing algorithm* and *tool-path generation algorithm*. This cross-domain security solution takes the feedback from the physical domain, and optimizes the cyber-domain design variables (*orientation* and *speed*) to minimize the mutual information between the cyber-domain G/M-codes and the physical domain analog emissions. For various benchmark 3D models, our solution obtains an average mutual information reduction of **24.94%** in acoustic side-channel, **32.91%** in power side-channel, **32.29%** in magnetic side-channel, and **55.65%** in vibration side-channel of the fused deposition modeling based Cartesian 3D printer. To validate the proposed methodology, a case study with an attack model is used to calculate the success rate. For the various threshold of G/M-code reconstruction, the average drop in success rate obtained is **8.74%**. With this chapter, the capability of a leakage-aware cyber-domain computer-aided manufacturing tool for maintaining the *confidentiality* of the system without physical system modification is highlighted.

References

1. Hojjati, A., Adhikari, A., et al. (2016) Leave your phone at the door: Side channels that reveal factory floor secrets. In *Proceedings of the 2016 ACM SIGSAC Conference on Computer and Communications Security*. New York: ACM.
2. Al Faruque, M. A., Chhetri, S. R., et al. (2016). Acoustic side-channel attacks on additive manufacturing systems. In *International Conference on Cyber-Physical Systems (ICCPS)*. Piscataway: IEEE.
3. Chhetri, S. R., et al. (2017). Confidentiality breach through acoustic side-channel in cyber-physical additive manufacturing systems. *ACM Transaction on Cyber-Physical Systems, 2*(1), 3
4. Yampolskiy, M., Andel, T. R., et al. (2014). Intellectual property protection in additive layer manufacturing: Requirements for secure outsourcing. In *Proceedings of the 4th Program Protection and Reverse Engineering Workshop*. New York: ACM.
5. Straub, J. (2017). Identifying positioning-based attacks against 3D printed objects and the 3D printing process. In *SPIE Defense+ Security*. International Society for Optics and Photonics.
6. Chhetri, S. R., Canedo, A., & Al Faruque, M. A. (2016). KCAD: Kinetic cyber-attack detection method for cyber-physical additive manufacturing systems. In *Proceedings of the 35th International Conference on Computer-Aided Design*. New York: ACM.
7. Belikovetsky, S., Solewicz, Y., et al. (2017). Detecting cyber-physical attacks in additive manufacturing using digital audio signing. arXiv preprint:1705.06454.
8. Tsoutsos, N. G., Gamil, H., & Maniatakos, M. (2017). Secure 3D printing: Reconstructing and validating solid geometries using toolpath reverse engineering. In *Proceedings of the 3rd ACM Workshop on Cyber-Physical System Security*. New York: ACM.
9. McMillen, D. (2016). *Security trends in the manufacturing industry targeting control systems and crown jewels*. Armonk: IBM. https://securityintelligence.com/media/security-trends-in-the-manufacturing-industry/.

10. Chen, F., Mac, G., & Gupta, N. (2017). Security features embedded in computer aided design (CAD) solid models for additive manufacturing. *Materials & Design, 128,* 182–194.
11. Gupta, N., Chen, F., et al. (2017). ObfusCADe: Obfuscating additive manufacturing CAD models against counterfeiting. In *Proceedings of the 54th Annual Design Automation Conference 2017.* New York: ACM.
12. Ashford, W. (2014). *21 percent of manufacturers hit by intellectual property theft.* Computer Weekly, [online] www.computerweekly.com.
13. Jamieson, R., & Hacker, H. (1995). Direct slicing of CAD models for rapid prototyping. *Rapid Prototyping Journal, 1*(2), 4–12.
14. Jin, G. Q., Li, W. D., et al. (2011). Adaptive tool-path generation of rapid prototyping for complex product models. *Journal of manufacturing systems, 30*(3), 154–164.
15. Yang, W., Zhou, Y., et al. (2017). Multi-channel fusion attacks. *IEEE Transactions on Information Forensics and Security, 12*(8), 1757–1771.
16. Mativo, T., Fritz, C., & Fidan, I. (2018). Cyber acoustic analysis of additively manufactured objects. *The International Journal of Advanced Manufacturing Technology, 96*(1–4), 581–586.
17. Veyrat-Charvillon, N., & Standaert, F.-X. (2009). Mutual information analysis: How, when and why?. In *CHES* (Vol. 5747, pp. 429–443). Berlin: Springer.
18. Chhetri, S. R., Faezi, S., & Al Faruque, M. A. (2017). Fix the leak! an information leakage aware secured cyber-physical manufacturing system. In *2017 Design, Automation and Test in Europe Conference and Exhibition (DATE).* Piscataway: IEEE.
19. Kramer, T. R., Proctor, F. M., et al. (2000). *The NIST RS274NGC interpreter-version 3* (Vol. 5416). NISTIR.
20. Ultimaker. (2017). *Cura software.* https://ultimaker.com/en/products/cura-software.
21. Drumm, B. (2011). Printrbot: Your first 3D printer. *Retrieved November, 26* (2013).
22. AT2021 (2016). *Cardioid Condenser Microphone.* Audio-Technica.
23. *Adafruit triple-axis accelerometer* (2017). www.adafruit.com.
24. Pico TA018 AC/DC Current Probe (2017). www.picotech.com.
25. *Magnetometer digital triple axis—HMC5883l* (2017). www.honeywell.com.
26. Thingiverse (2017). https://www.thingiverse.com/.

Chapter 5
Data-Driven Kinetic Cyber-Attack Detection

5.1 Introduction

In Chap. 4 we discussed a data-driven modeling approach to reduce the information leakage from the side-channels. In this chapter, we will present a data-driven modeling approach for providing defense against a new kind of attacks highly applicable to cyber-physical systems. This kind of attacks is known as kinetic cyber-attacks. These type of attacks originate from the cyber-domain but cause physical damage, injury, or even death [1]. In CPS, effects of kinetic cyber-attacks have been recently highlighted by incidents such as the Stuxnet malware [2], Maroochy water breach [3], German steel mill cyber-attack [4], and security breaches in automobiles [5]. In cyber-physical additive manufacturing systems, kinetic cyber-attack can find its way through the digital process chain to introduce various inconspicuous flaws in the 3D objects. If these objects are critical for the system, they can compromise the structural integrity and pose a severe safety risk. For example, researchers placed an inconspicuous void (which was less than 1 mm in dimension) in the 3D design of a standard D638-10 tensile specimen provided by American Society for Testing and Materials (ASTM). This void reduced the mechanical strength of the specimen to carry load by 14% [6].

The security concerns in CPS are not new [7, 8], and in fact, various attack detection methods have been designed for the identification and detection of attacks [9]. However, cyber-physical additive manufacturing system has not received much attention for the research in attack detection methods. In [6], authors present the potential attack vectors for cyber-physical additive manufacturing systems' digital process chain and recommend securing the process chain by incorporating software checks, hashing, and process monitoring through side-channel. In [10], authors present a signature-based attack detection method leveraging the concept of integrated circuit Trojan detection from side-channel analysis and system health monitoring. However, it should be noted that signature-based technique requires

© Springer Nature Switzerland AG 2020
S. R. Chhetri, M. A. Al Faruque, *Data-Driven Modeling of Cyber-Physical Systems Using Side-Channel Analysis*,
https://doi.org/10.1007/978-3-030-37962-9_5

acquiring the signature of a baseline structure every time it is created, which is counter-intuitive for rapid prototyping capability of the cyber-physical additive manufacturing system. In [11], authors list the threat surface and describe its effects on the manufacturing parameters. In the cyber-physical additive manufacturing system, any variation made in the design of the 3D object is manifested physically. Moreover, during the printing stage, the machine itself unintentionally emits *analog emissions* from the side-channels (such as acoustics, electromagnetic radiation, power, etc.). Thus, we propose a novel kinetic cyber-attack detection (KCAD) method that uses statistical data-driven modeling of the cyber-physical additive manufacturing system to detect the anomalous *analog emission* which can arise as a result of potential *zero-day kinetic cyber-attacks*.

5.1.1 Motivation

The fundamental motivation for KCAD method comes from the fact that in CPS, the information flow in the cyber-domain has at least one corresponding signal flow in the physical domain [12]. These signals in the physical domain actuate the physical processes, and this actuation converts energy from one form to the another. This phenomenon allows us to monitor the unintentionally leaked *analog emissions* which have high *mutual information* with the corresponding control signals. *Analog emissions* have been used in system health monitoring and prognostics to infer about the current state of the system, and also for quality control in manufacturing [13, 14]. However, traditional quality control system focuses on measuring the key quality characteristics, and kinetic cyber-attack on various features may not be detected by such systems [10]. On the other hand, these *analog emissions* are also used to breach the confidentiality of the system [15]. However, incorporating statistical method, the acquired *analog emissions* from the side-channel corresponding to the *control signals* can be used to model the behavior of the system. And this model may be used for detecting the intrusion in the system [16].

5.1.2 Problem and Challenges

The current challenge for securing the CPS is understanding its unique properties and vulnerabilities and preventing the possibility of an attack [17]. However, *zero-day vulnerabilities* of the system are hard to detect during design-time. In such scenarios, detection of such attacks is the best possible defense. After detection, we can take measures such as halting the system, perform corrective actions, etc., to mitigate the effects of the attack. Details of the mitigation methods are out of the scope of this chapter. In the cyber-physical additive manufacturing system, the problem for designing a *zero-day kinetic cyber-attack* detection method poses the following key challenges:

1. Detecting the intrusion/attack that can occur at various points of the digital process chain of cyber-physical additive manufacturing system, and affect the dynamics of the system.
2. Making it non-intrusive so that it can be used in legacy cyber-physical additive manufacturing system systems.
3. Complementing the detection method with the physical and process knowledge from cyber-physical additive manufacturing system.

5.1.3 Contributions

To address the above-mentioned challenges, we propose a novel attack detection method [18] for detecting the effects of possible *kinetic attacks* in the digital process chain of the additive manufacturing that employs:

1. **Modeling of an Adversary (Sect. 5.2)** to understand the various attack points in the digital process chain for effective implementation of the attack detection method.
2. **Statistical Estimation (Sect. 5.3)** to perform data-driven modeling of the behavior of the cyber-physical additive manufacturing system by analyzing the relationship between the *analog emissions* and the control signals.
3. **Analysis of Analog Emission (Sect. 5.3.3)** to use it as a parameter from the side-channel for estimating the relationship between the *analog emissions* and the control parameters using **mutual information** as the relation measurement metric.

5.2 Kinetic Cyber-Attack Adversary Model

In the cyber-physical additive manufacturing system, information (3D design specification) flows through the digital process chain and is finally converted as the control signals in the machine. In the adversary model, an attacker can infiltrate at various stages of the process chain (see Fig. 5.1) to alter the *integrity* of the tools, algorithms, and firmware. Moreover, they may add exogenous inputs e_1, e_2, and e_3 during the transfer of information from one stage to the other in the digital process chain.

Let us consider y to be the control signals to the physical components of the cyber-physical additive manufacturing system machine, provided ψ is the true information. In this case, $\tilde{u}'$, $\tilde{u}''$, and $\tilde{y}'''$ are the false information produced in the process chain due to the compromised *integrity* of the CAD tool, slicing algorithm, and the firmware, respectively. We can observe that the attacks on the digital process chain by different attackers A_1, A_2, and A_3 always result in the modification of the control signals y. This phenomenon is what separates the CPS from traditional information and technology systems. In FDM based cyber-physical

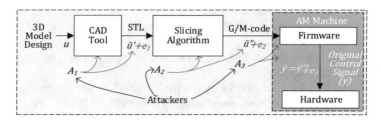

Fig. 5.1 Adversary attack points in the digital process chain

additive manufacturing system, the outcome of kinetic attack will result in the variation of control signal y to $\tilde{y}$, such that it modifies the initial 3D design of the object.

Remark 5.1 In FDM based cyber-physical additive manufacturing system, the change in the information flow by an attacker will result in the change of control parameters $y = [v, a, t, d]$ responsible for controlling the dynamics of the machine.

where $v = [v_x, v_y, v_z, v_e]$ represents the speed of the nozzle in different axis, along with the speed of extrusion, and $v_{i \in \{x,y,z,e\}} \mathbb{R}_{\geqslant 0}$. $a = [a_x, a_y, a_z, a_{xy}, a_{xz}, a_{yz}, a_{xyz}]$ represents the axis of movement such that $a_{i \in \{x,y,z,xy,xz,yz,xyz\}} \in \{0, 1\}$, and $a = 1$ represents presence of movement in the given axis, and $a = 0$ represents absence of movement. $t \in \mathbb{R}_{\geqslant 0}$ represents the temperature of the nozzle. $d = [d_x, d_y, d_z, d_e]$ represents the distance of the nozzle in different axis, along with the amount of extrusion, and $d_{i \in \{x,y,z,e\}} \in \mathbb{R}$. Hence, the kinetic cyber-attack in FDM based cyber-physical additive manufacturing system will cause the information u to be altered such that the final control parameters to the physical components are altered to $\tilde{y} = [\tilde{v}, \tilde{a}, \tilde{t}, \tilde{d}]$.

Definition With reference to Remark 5.1, the kinetic cyber-attack in FDM can be defined as change in the information flow u in the digital process chain such that:

$$
\begin{bmatrix} v \\ a \\ d \\ t \end{bmatrix} \pm \begin{bmatrix} e_v \\ e_a \\ e_d \\ e_t \end{bmatrix} = \begin{bmatrix} \tilde{v} \\ \tilde{a} \\ \tilde{d} \\ \tilde{t} \end{bmatrix} \tag{5.1}
$$

and $\tilde{y} \neq y$, when $\sum_{i \in \{v,a,d,t\}} e_i > 0$. Here $\{e_v, e_d, e_t\} \in \mathbb{R}_{\geqslant 0}, e_a \in \{0, 1\}$.

5.3 KCAD Method

Let $Y \to O$ be a side-channel, where Y and O represent random variables denoting control information parameters and observed *analog emission*, respectively. Then, KCAD method leverages the fact that these variables have high mutual information.

5.3.1 Mutual Information

Remark 5.2 Let the observed *analog emissions* be $o(t)$, then the control parameters, $y = [v, a, t, d]$, responsible for controlling the dynamics of the system, emit *analog emissions* such that the mutual information $\{I(V; O), I(A; O), I(T; O), I(D; O)\}$ > 0, where (V, A, T, D) are random variables.

The random variables O, V, T, and D are continuous whereas A is discrete. Let $f(o)$, $f(v)$, $f(t)$, and $f(d)$ be probability density function (*pdf*) of continuous random variables O, V, T, and D, respectively. And for all $k, j \in \{o, v, t, d\}$, let $f_{k \neq j}(k, j) = f_{k \neq j}(j, k)$ be the joint *pdf*. The conditional *pdf* for these random variables is then defined as $f_{k \neq j}(k|j) = \frac{f_{k \neq j}(k, j)}{f(j)}$. Then, we can calculate the differential entropy of these random variables as follows:

$$h_{K \in \{O, V, T, D\}}(K) = -\int f(k) \log(f(k)) dk \qquad (5.2)$$

Similarly, the conditional differential entropy of these random variables can be given as follows:

$$h_{K, J \in \{O, V, T, D\}}(K|J)_{K \neq J} = -\int \int f(k, j) \log(f(k|j)) dk dj \qquad (5.3)$$

The mutual information between the observed *analog emission* and the control parameters can be given as follows:

$$I_{K \in \{V, T, D\}}(K; O) = h(K) - h(K|O) \qquad (5.4)$$

Let $f(a)$ be a probability distribution function of the discrete random variable A. Then the entropy of A can be calculated as follows:

$$H(A) = -\sum f(a) \log(f(a)) \qquad (5.5)$$

For calculating the mutual information between O and A, we can divide the values of O into bins of length ε. If $H(O_\varepsilon)$ be the entropy of O after discretization, we have $h(O) = \lim_{\varepsilon \to 0} [H(O_\varepsilon) + \log(\varepsilon)]$. And the mutual information can be calculated as:

$$I(A; O) = H(A) - H(A|O_\varepsilon) \qquad (5.6)$$

The calculation of mutual information between two continuous random variables requires estimation of the probability density functions, which are then used in Eqs. 5.2 and 5.3. Kernel probability density estimation can be used for estimating the *pdf* based on the experimental data as:

$$\tilde{f}(x) = \frac{1}{nh} \sum_{i=1}^{n} K\left(\frac{x - x_i}{h}\right) \tag{5.7}$$

where K is a real-valued integrable kernel function and h the bandwidth. There are various kernel function that can be used for the estimation.

Our proposed KCAD method can be defined as a pipeline of deterministic algorithms which takes continuous observable *analog emissions* $o(t)$ as input along with the information flow U in the form of G-codes $G_t = [g_1, g_2, g_3, \ldots, g_t]$ from which the control parameters v, a, d, and t are parsed. The information (U) acquired by KCAD is assumed to be from a secure channel and free from any modification. KCAD method infers the *analog emission* O based on the given control parameters. Given the presence of a kinetic cyber-attack, the control parameters are changed to $\tilde{y} = [\tilde{v}, \tilde{a}, \tilde{d}, \tilde{t}]$ with observed *analog emissions* being $\tilde{O}$, such that $|\tilde{O} - O| = e$. And for $e > e^T$, emission variation threshold, the output of the detection system is *True*, denoting presence of an attack.

Definition Attack Detectability: Given the input O and G with parsed variables v, a, d, and t, kinetic attack to the system, such that $\sum_{i \in \{v,a,d,t\}} e_i > 0$, is detected by KCAD method if its output is true.

5.3.2   KCAD Architecture

The architecture of the simplified KCAD method is shown in Fig. 5.2. One of the advantages of this method is that it can be placed to monitor the information flow in any of the stages of the digital process chain. With this, the attack on the integrity of any tools, firmware, and algorithms can be detected if it has the corresponding effect on the dynamics of the system. KCAD runs in parallel to the cyber-physical additive manufacturing system while it is printing, and non-intrusively and continuously acquires the observable *analog emissions*. The components of KCAD that work in parallel to the cyber-physical additive manufacturing system depends on which point of the process chain we feed the information to it. For example, if we want to detect the attack on the integrity of the firmware of the cyber-physical additive manufacturing system, the input to the KCAD can just be the G-code/M-code. The slicing algorithm in it can thus be switched off. However, the channel through which the information passes to the KCAD method from different point of the digital process chain is assumed to be secure. This means that the KCAD method will always receive original/unmodified cyber data from the digital process chain.

Analog Emission Sensors: Various sensors (piezoelectric, current, electromagnetic, etc.) can be used to monitor the *analog emissions* from the cyber-physical additive manufacturing system. Given the integrity attack that introduces values e_v, e_a, e_d, and e_t, in the control parameters (v, a, d, t), the sampling frequency (F_s) and bandwidth (B) should be such that it can measure corresponding changes e_v, e_a,

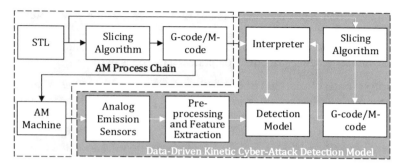

Fig. 5.2 Architecture of data-driven kinetic cyber-attack detection method

e_d, and e_t in the *analog emissions* $o(t)$. Moreover, distance and angle of placement of the sensors also affect the signal-to-noise ratio (SNR) given as:

$$\text{SNR}_{dB} = 10 \log_{10} \left(\frac{P_{Signal}}{P_{Noise}} \right) \tag{5.8}$$

Hence, the choice and placement of the sensors depends on the choice of side-channel and the relation between the *analog emissions* and the control parameters. For choice of side-channel and the corresponding *analog emissions*, Eqs. 5.4 and 5.6 can be used as a measure of relation between the observable *analog emissions* and the control parameters. Based on this measurement, the observed values can either be incorporated or discarded from group of features to be used for estimating the behavior of the system in KCAD.

Pre-processing and Feature Extraction: Pre-processing is done to improve the SNR by removing the known noise signals from the *analog emissions* that are independent of the control parameters. If observable *analog emission*, $o(t) \neq f(y(t))$, where y represents the control parameters, then the observed *analog emission* can be considered as noise. The signals acquired by sensors can be too large for the estimation algorithms used in the detection model. Hence, it is necessary to extract only the informative values from the signal to improve the processing time of the detection model. For each observed signal various values (features) are derived using the original signal.

$$O_t = \begin{bmatrix} o_1^{f^1} & o_1^{f^2} & o_1^{f^3} & \dots & o_1^{f^n} \\ o_2^{f^1} & o_2^{f^2} & o_2^{f^3} & \dots & o_2^{f^n} \\ \vdots & \vdots & \vdots & \ddots & \vdots \\ o_t^{f^1} & o_t^{f^2} & o_t^{f^3} & \dots & o_t^{f^n} \end{bmatrix} \tag{5.9}$$

where the column represents the number of features and the row represents the discretized values of the signal $o(t)$. The type of features extracted is specific to the observed *analog emissions* (for acoustics see Sect. 4.2). For reducing the dimension of the extracted features, principal component analysis is used.

Interpreter: Each block (line) of instruction (G/M-code) sent to the cyber-physical additive manufacturing system consists of control parameters. These instructions or numerical control codes are converted to the canonical machining commands using interpreters such as NIST RS274NGC. In our KCAD method, a lighter version of Arduino G-code and NIST RS274NGC interpreter is used to extract the control signals v, a, d, and t. These signals are then sent to the detection model.

Detection Model: The detection model uses the supervised learning approach to estimate the function $\hat{f}_i(O_t, \alpha_n)$ with $i = 1, 2, 3, 4$ using the training dataset of observed *analog emissions* (see Fig. 5.3). For each control parameters we estimate the function with respective parameter α. Various predictive models can be estimated based on the initial training datasets. The training datasets will require to balance the trade-off between bias and variance to find optimal parameter α. k-fold cross-validation is used to perform data driven validation for the estimated function $\hat{f}_n(.)$. For the regression function estimation, learning algorithms such as *gradient boosting regressor (GBR)*, *ridge regression*, *stochastic gradient descent regression (SGDR)*, *Bayesian ridge regression (BRidge)*, *passive aggressive regression (PAR)*, *decision tree regression (DTR)*, *elastic net regression (ENet)*, *linear regression with lasso*, and *k-nearest neighbor regression (kNN)* is used. Each of these models is compared using metrics such as *explained variance*, *mean absolute error (MAE)*, *mean squared error (MSE)*, *median absolute error*, and R^2 *score*. For function that needs to be estimated for classification, classifiers such as *support vector machine (SVC)* with *linear* and *radial basis function (RBF)* as kernels, *logistic regression*, *stochastic gradient descent classifiers*, *ensemble of AdaBoost* and *gradient boosting* are used. They are compared using receiver operating characteristic (ROC) curves, and the model with the least error is selected.

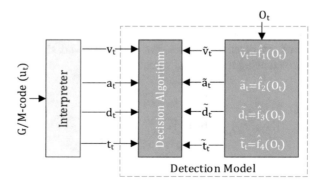

Fig. 5.3 Data-driven detection model

Algorithm 1: Detection algorithm

    Input: Real and Estimated Control Parameters $[v, a, d, t]$, $[\tilde{v}, \tilde{a}, \tilde{d}, \tilde{t}]$
    Output: Attack Flag F_A
1 Define Error Thresholds $e_v^T, e_a^T, e_d^T, e_t^T$
2 Initialize $f_v = 0, f_a = 0, f_d = 0, f_t = 0$
3 **for** $each\ i \in v, a, d, t$ **do**
4   **if** $|i - \tilde{i}| > e_i^T$ **then**
5          $f_i = 1$

6 **if** $\sum_{i \in \{v,a,d,t\}} f_i \geq 1$ **then**
7       $F_A = 1$

8 **return** F_A

Detection Algorithm: The detection algorithm compares the real control parameters given by the interpreter and the values calculated by the estimated functions $\hat{f}_n(.)$. In Algorithm 1, in Line 1, the error variation thresholds are defined. This value is based on the accuracy of the estimated functions $\hat{f}_n(.)$ during the training phase of the detection model. Lines 3–5 determine if the estimated control parameters and the real control parameters vary more than the error threshold. Lines 6–7 set the attack detection flag to high if any of the control parameters varies more than the error variation threshold. Finally in Line 8 the attack flag is returned.

Offline vs Online Model Estimation The model function $\hat{f}_n$ estimation can be done either online or offline. This is necessary to consider because of the fact that cyber-physical additive manufacturing system machine will have varying observable *analog emission* over a long period of time due to wear and tear of the mechanical structures. Offline function estimation can have a shorter response time. However, an online estimation can be done for higher accuracy with a longer response time.

5.3.3 Acoustic Analog Emissions

Acoustic *analog emissions* is one of the observable emissions in the cyber-physical additive manufacturing system. The fundamental working principle behind KCAD method is that the *analog emissions* $o(t)$ must have high mutual information with the control parameters $y(t)$. As a proof of concept of the KCAD method, we will use acoustics as the observable *analog emissions* to detect the presence of kinetic cyber-attack on the cyber-physical additive manufacturing system. However, it is trivial to infer that the observed *analog emission* will have a weak relation with the control parameter temperature t. Therefore, only a kinetic cyber-attack affecting the control parameters v, a, and d will be considered. The main source of acoustics in FDM based 3D printers is the vibration of the stepper motors. These 3D printers consist of at least one stepper motor to control the movement of the nozzle of the printer in each axis (x, y, and z axes) [19]. These stepper motors consist of rotor (permanent magnet) and the stator (electromagnet). The varying radial electromagnetic force acting on the stator of the stepper motor produces vibration [20–22]. This vibration is the source of acoustics. The radial electromagnetic force is controlled by the control parameters, such as *speed* of the movement of the motor. However, the natural frequency of the stator is determined by the load, connected frame, and the structure of the stator [23]. Hence, resonance occurs when the vibration produced by the radial electromagnetic matches the harmonics of the natural frequency of the stator. This resonance frequency is different for different stepper motors responsible for moving the 3D printer's nozzle in x, y, and z. This will allow us to estimate functions to separate the movement in different axes. The *analog emission* sensors for capturing the acoustic emissions will require sampling frequency of more than 40 kHz to capture the range of audible sound 20 Hz to 20 kHz. During the pre-processing stage digital filter is used to remove low and high frequency noise. Dynamic time warping is used to dynamically assign the window size w for feature extraction. However, an initial fixed length window size (10–50 ms) will be used to extract features size as *zero crossing rate*, *energy entropy*, *spectral entropy*, and *Mel frequency cepstral coefficients (MFCCs)*. Using these features, euclidean distance is measured to define the dynamic window size for accurate feature extraction. As shown in Algorithm 2, Line 4 measures the euclidean distance between the features of the previous and current *analog emission*. If this distance is greater than the threshold distance d^T (which is determined by utilizing the training dataset), then only the window size gets changed. As windowing is done to extract the features from the observed *analog emissions* $o(t)$, we can determine the control parameter v from d and vice versa, where $v = d/w$, where w is length of the window in seconds.

Remark 5.3 For FDM based cyber-physical additive manufacturing system, KCAD method will be able to monitor any kinetic attacks modifying the control parameters v, a, and d by analyzing the variation in the acoustic *analog emissions* to the corresponding control parameters v and a.

Algorithm 2: Dynamic window size determination

Input: Observed Analog Emission $o(t)$
Output: Dynamic Windows $w = [w_1, w_2, \ldots, w_n]$
1 Initialize $n = 1$, $i_{previous} = 1$
2 Extract Features O_t // $n \rightarrow$ Number of Features
3 **for** $i = 2$ *to* t **do**
4   **if** $|\sqrt{(o_i^{f^1} - o_{i-1}^{f^1})^2 + \ldots + (o_i^{f^n} - o_{i-1}^{f^n})^2}| > d^T$ **then**
                                                                    // $d^T \rightarrow$ Threshold Distance
5          $w_n = i - i_{previous} + 1$
6          $i_{previous} = i + 1$
7          $n = n + 1$
8 **return** w

5.3.4 Performance Metrics

The performance of an attack detection method can be measured with two metrics *true positive rate (TPR)* and *true negative rate (TNR)*.

$$TPR = \frac{TP}{TP + FN} \qquad (5.10)$$

where *true positive* (TP) is the total number of positive detection when there is an attack in the system, and *false negative* (FN) is the total number of negative detection during the presence of an attack. Similarly,

$$TNR = \frac{TN}{TN + FP} \qquad (5.11)$$

where *true negative* (TN) is the total number of positive detection when there is not any attack to the system, and *false positive* (FP) is the total number of positive detection when there is not any attack to the system. Then, the accuracy of the system can be measured as follows:

$$\text{Accuracy} = \frac{TP + TN}{\text{Total sample}} \qquad (5.12)$$

5.4 Experimental Results

5.4.1 Experimental Setup

The experimental setup for the KCAD method evaluation is shown in Fig. 5.4. This evaluation tests KCAD performance against integrity attacks on the printer's firmware. In a simple scenario, a modified firmware is installed in the 3D printer by an attacker. This attack modifies the control signal to the 3D printer, which introduces variation in the geometry of the 3D object. The audio recorder is placed

Fig. 5.4 KCAD method experimental setup

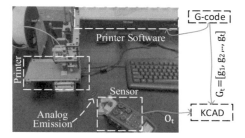

at an optimal distance to acquire the acoustics. In the experiment, the recorder is placed at $45°$ angle to the x and y axes to acquire the variation of the signal in both directions with a single recorder. During the training phase, G-codes are written to move the nozzle of the printer in various x and y axes directions with various printing speeds (400–4500 mm/min with 100 mm/min step size). These speed ranges are machine-specific. Using this training data, we estimated the model function for the control parameters v_x, v_y, and $a = [a_x, a_y, a_z, a_{xy}]$. Using a similar approach, model functions for parameters v_e, v_z, different directions, and $a = [a_{yz}, a_{xz}, a_{xyz}]$ can be estimated.

5.4.2   Mutual Information Calculation

To demonstrate the dependency of the acoustic *analog emissions* with the control parameters $v = [v_x, v_y, v_z]$, and the $a = [a_x, a_y, a_z, a_{xy}]$ mutual information between them is calculated. For control parameters a, it is treated as a discrete random variable with different labels depending on the combination of axis movements. Table 5.1 shows the mutual information calculated for the control parameters and the features extracted from the observed *analog emissions*. We can see that different features have varying mutual information with the control parameters. Also, it can be observed that the mutual information between the speed in $z\ axis$ and the *analog emission* is comparatively low. This is due to the reason that speed in $z\ axis$ is almost constant in most of the 3D printers. The estimation function utilizes different features on the basis of the mutual information and principal component analysis to select the features that are most relevant.

5.4.3   Model Function Estimation

The function, $\hat{f}_i(O_t, \alpha_n)$, estimation is the fundamental step in our KCAD method. The parameter $[\alpha_1, \alpha_2, \ldots, \alpha_n]$ are responsible for minimizing the cost functions

Table 5.1 Mutual information in bits between features and control parameters

	Features													
	$o_t^{f^1}$	$o_t^{f^2}$	$o_t^{f^3}$	$o_t^{f^4}$	$o_t^{f^5}$	$o_t^{f^6}$	$o_t^{f^7}$	$o_t^{f^8}$	$o_t^{f^9}$	$o_t^{f^{10}}$	$o_t^{f^{11}}$	$o_t^{f^{12}}$	$o_t^{f^{13}}$	$o_t^{f^{14}}$
v_x	0.79	0.95	0.16	0.81	0.74	0.76	0.35	1.18	1.51	0.31	0.55	0.69	0.75	0.71
v_y	0.27	0.67	0.11	0.43	0.41	0.24	0.23	1.31	1.10	0.23	0.44	0.45	0.55	0.60
v_z	**0.16**	**0.07**	**0.0806**	**0.08**	**0.07**	**0.08**	**0.07**	**0.07**	**0.07**	**0.07**	**0.07**	**0.07**	**0.08**	**0.09**
a	0.58	0.82	0.18	0.69	0.69	0.53	0.44	1.91	1.17	0.64	0.74	0.22	0.40	0.39

The bold values demonstrate that mutual information between the z axis speed variation and the feature is the lowest.

used by the learning algorithms. This estimation is done in the training phase. Based on the relation between the control parameters and the features extracted from observed *analog emissions*, the estimated functions can be used for regression or classification. The relation between the control parameter v and observed *analog emissions* $o(t)$ both being continuous random variable can be estimated using regression algorithms, whereas, the parameter a is a discrete random variable and we have to use classifiers for estimating the function $\hat{f}(.)$. We will use various algorithms and perform the comparison between them to select the function $\hat{f}(.)$ that gives the least error.

Regression: Function $\hat{f}_{[v_x, v_y]}(O_t, \alpha_n)$ is estimated for the control parameter v based on O_t. From Table 5.2, it is clear that *gradient boosting regression* outperforms the rest of the regression models in terms of the error metrics. Hence, it is selected to estimate the function for the control parameters (v_x, v_y). Function for v_z is not estimated as speed in z axis is constant.

Classification: For the parameter a, various functions estimated along with their performance is compared in Fig. 5.5. The estimation function is treated as one versus the rest classification and the ROC is calculated as an average of ROC of all the classes, i.e., various movement axes, $[a_x, a_y, a_z, a_{xy}]$. From Fig. 5.5, *logistic regression* classifier is selected as the optimal function for the classification. In Fig. 5.6, the ROC curve of various classes calculated as a one versus rest classification using *Logistic Regression* classifier is presented. Since the area under the curve for the z $axis$ movement class is higher, the detection of the presence of nozzle movement in z $axis$ is easiest compared to the other classes. With these functions estimated for various control parameters, KCAD can effectively detect the variation in the *analog emissions* with the expected emissions in the presence of kinetic cyber-attacks on the firmware of the 3D printer.

Table 5.2 Regression models comparison for $\hat{f}_v(.)$

Model	MSE	Variance	MAE	Median AE	R^2
GBR	0.0076	0.9923	0.0037	0.0167	0.9923
Ridge	0.0147	0.9854	0.0090	0.0775	0.9851
SGDR	0.0148	0.9852	0.0090	0.0785	0.9850
BRidge	0.0149	0.9852	0.0089	0.0772	0.9849
PAR	0.0183	0.9817	0.0095	0.0899	0.9815
DTR	0.0258	0.9741	0.0090	0.0582	0.9740
ENet	0.0786	0.9212	0.0210	0.1990	0.9208
Lasso	0.1015	0.8980	0.0241	0.2331	0.8976
kNN	0.0025	0.4997	0.0014	0.0182	0.4997

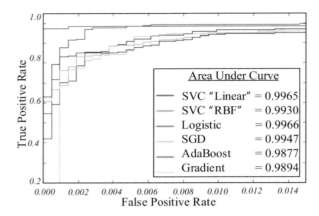

Fig. 5.5 Receiver operating characteristics curve of classifiers

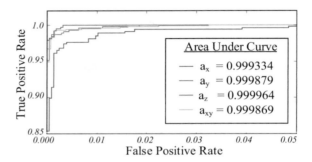

Fig. 5.6 Receiver operating characteristics curve of classification with one versus rest

5.4.4   Results for Detection of Kinetic Attack

We developed a zero-day kinetic attack on the 3D printer's firmware to test the
KCAD method. Our attack modifies the nozzle speed in the x and y direction while
printing, thus effectively changing the dimension of the object. Additionally, the
axis values are changed resulting in the deformation of the object. For detecting
the variation of speed on $x\ axis$ and $y\ axis$, the speed is varied from 600 to
3500 mm/min, while the attack is assumed to introduce range of variation in the
original speed, i.e., from 50 to 1000 mm/min. From the function estimation, error
threshold for speed e_v^T is set as 200 mm/min.

Tables 5.3 and 5.4 show that for higher speed and lower speed variation (δv),
the true positives are lower compared to the low speed and high speed variation.
However, the false positives are higher for higher speeds.

Intuitively, this is due to the fact that feature extracted prominently consists of
MFCC, which focuses on extracting more features from lower frequency range

Table 5.3 True positives for speed variation

δv	TP for speed (*100 mm/min)									Total
mm/min	7	9	12	15	17	20	25	30	35	TP
1000	16	16	16	16	16	16	16	16	16	144
500	16	16	16	16	16	14	15	16	14	139
300	16	16	16	13	10	7	7	8	7	92
200	13	16	10	9	8	6	8	8	6	84
TPR										0.7968

Table 5.4 False positives for speed variation

δv	FP for speed (*100 mm/min)									Total
mm/min	7	9	12	15	17	20	25	30	35	FP
1000	5	4	3	6	5	7	7	6	8	51
500	4	3	2	5	5	8	6	8	8	54
300	2	2	3	5	5	7	6	6	7	47
200	3	1	3	5	2	8	7	8	7	44
FPR										0.3402

Table 5.5 True positives for distance variation

δd	TP for speed (*100 mm/min)									Total
mm	7	9	12	15	17	20	25	30	35	TP
20	16	16	16	16	16	16	16	14	16	142
10	16	15	16	16	16	15	14	13	14	135
5	16	13	14	10	12	14	13	12	15	119
3	14	10	12	10	11	10	12	11	13	103
TPR										0.8663

rather than higher frequencies causing poor function estimation for higher speeds. KCAD accuracy for detection of attack on control parameter v can be calculated using Eq. 5.12 as 72.83%.

For testing the performance of KCAD method for detecting the change in the control parameter d introduced by the modified firmware, distance is varied from 3 to 20 mm. The error threshold for the distance is selected as 3 mm based on the error in the estimated function.

Table 5.5 shows that the number of true positive is decreasing with the increasing speed and lower distance variation δd. Moreover, Table 5.6 shows that the number of false positive is increasing with the increasing speed. Using Eq. 5.12, the accuracy for detection of attack on control parameter d is calculated to be 79.25%. We modified the movement in x, y, z, and xy axes to measure the KCAD performance against firmware modification attacks that vary the control parameter a, which determine the movement axis. Tables 5.7 and 5.8 shows that the true positive rate is decreasing with increase in speed and false positive rate is increasing with increase in the speed. The accuracy for detection of attack on control parameter a is 79.07%.

Table 5.6 False positives for distance variation

δd	FP for speed (*100 mm/min)									Total
mm	7	9	12	15	17	20	25	30	35	FP
20	4	4	4	6	4	3	3	5	4	37
10	3	4	4	5	4	4	4	5	5	38
5	2	5	5	5	5	4	5	4	7	42
3	2	4	7	7	5	2	5	5	8	45
FPR										0.2812

Table 5.7 True positives for axis variation

$Axis, Total$	TP for speed (*100 mm/min)									Total
TP	7	9	12	15	17	20	25	30	35	TP
$a_x, 32$	32	32	28	28	24	21	21	20	19	225
$a_y, 32$	31	32	27	25	23	19	19	19	18	213
$a_{xy}, 24$	20	21	21	19	19	20	19	18	17	174
$a_z, 24$	24	20	19	18	18	19	18	19	18	173
TPR										0.7787

Table 5.8 False positives for axis variation

$Axis, Total$	FP for Speed (*100 mm/min)									Total
TN	7	9	12	15	17	20	25	30	35	FP
$a_x, 32$	2	2	2	3	4	6	6	7	9	41
$a_y, 32$	2	3	4	2	4	5	4	9	8	41
$a_{xy}, 24$	4	4	6	6	7	5	6	7	8	53
$a_z, 24$	5	6	7	8	7	8	7	8	8	64
FPR										0.1974

Our KCAD method relies on the fact that any *zero-day kinetic cyber-attack* results in variation of control parameters v, a, and d. Hence, treating these parameters as synthetic benchmarks, the accuracy of the KCAD method for measuring the variation of various control parameters v, a, and d is 77.45%.

5.4.5 *Test Case: Base Plate of a Quad Copter*

As a test case, we present an analysis on a flight controller base plate [24], which is a part of the quadcopter that can be printed using a 3D printer. We assume that the modified firmware, as a result of *zero-day kinetic cyber-attack*, introduces variation in the certain part of the code by adding 4 mm to the *x axis* movement distance. Such small changes in the design of an object can compromise their structural integrity during use, and lead to a catastrophic failure. The original design of the base plate is shown in Fig. 5.7a. As a result of an attack, minute modification

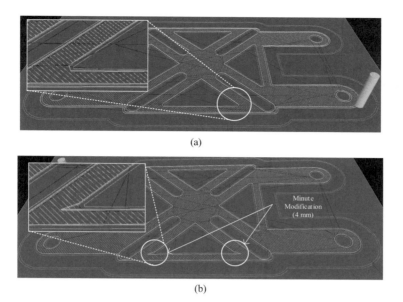

(a)

(b)

Fig. 5.7 Attack on base plate of a quad copter. (**a**) Original G-code trace. (**b**) G-code trace after kinetic attack

introduced in the design is shown in Fig. 5.7b. This modification might not be visible to human eyes; however, it compromises the structural integrity of the base plate. KCAD method detected the variation in the x axis introduced in all three layers by the firmware modification attack.

5.5 Discussion

In this section, we provide an open discussion on the limitations of our KCAD method:

1. **Printer Variation:** KCAD is machine-specific. Hence, for implementation, the function estimation (training) has to be conducted before it can be implemented. Different FDM based 3D printers will emit different acoustic signature based on the type of motors used and the structure of the frame. This has to be studied before KCAD can be implemented. Different *analog emissions* and their respective features have to be analyzed for model function estimation. However, this has to be done only once before the implementation.
2. **Complex Attacks:** In the experimental section, we have tested the KCAD with firmware modification attacks introducing simple variation in the x, y axes. However, more complex variation can be introduced by the attack, with attacks modifying both the x and y axes together. In such scenarios, more function has to be estimated for a combination of control parameters.

3. **Sensor Placement:** The *analog emissions* sensors placement should not obstruct the printing process. However, it must be able to capture the emissions with high SNR. This requires analyzing various sensors and their optimal position. It might have to be placed inside the system for better SNR. However, this can only be done if it does not obstruct the printing process.

5.6   Summary

In this chapter, we presented a novel data-driven kinetic cyber-attack detection method, which can be placed non-intrusively and can monitor the system during run-time. We have performed the analysis of acoustic *analog emissions* to measure the feasibility of such a system, and demonstrated that acoustics have high mutual information with the control parameters parsed from the cyber-domain data. We have tested our system on an FDM based 3D printer, assuming that integrity attack on the printer firmware eventually modifies the control parameters to the physical components. We have tested the performance of our method with a variety of speed, distance, and axis that can be an outcome of the kinetic cyber-attack in the digital process chain of the additive manufacturing. KCAD method achieves a high accuracy of 77.45%. This provides a good starting point and proof of concept for the proposed attack detection method.

References

1. Applegate, S. D. (2013). The dawn of kinetic cyber. In *2013 5th International Conference on Cyber Conflict (CyCon)* (pp. 1–15) . New York: IEEE.
2. Falliere, N., Murchu, L. O., & Chien, E. (2011). *W32. stuxnet dossier.* White paper, Symantec Corp., Security Response, 5.
3. Slay, J., & Miller, M. (2007). *Lessons learned from the Maroochy Water Breach.* New York: Springer.
4. Lee, R. M., Assante, M. J., & Conway, T. (2014). *German steel mill cyber attack.* Industrial Control Systems, 30.
5. Koscher, K., Czeskis, A., Roesner, F., Patel, S., Kohno, T., Checkoway, S., et al. (2010). Experimental security analysis of a modern automobile. In *2010 IEEE Symposium on Security and Privacy (SP)* (pp. 447–462). Piscataway: IEEE.
6. Sturm, L., Williams, C., Camelio, J., et al. (2014). Cyber-physical vulnerabilities in additive manufacturing systems. In *Context, 2014*.
7. Al Faruque, M., Regazzoni, F., & Pajic, M. (2015). Design methodologies for securing cyber-physical systems. In *Proceedings of the 10th International Conference on Hardware/Software Codesign and System Synthesis*. New York: IEEE Press.
8. Wan, J., et al. (2016). Exploiting wireless channel randomness to generate keys for automotive cyber-physical system security. In *2016 ACM/IEEE 7th International Conference on Cyber-Physical Systems (ICCPS)* (pp. 1–10). New York: IEEE.
9. Mitchell, R., & Chen, I.-R. (2014). A survey of intrusion detection techniques for cyber-physical systems. *ACM Computing Surveys (CSUR), 46*(4). Article No. 55.

10. Vincent, H., Wells, L., Tarazaga, P., & Camelio, J. (2015). Trojan detection and side-channel analyses for cyber-security in cyber-physical manufacturing systems. *Procedia Manufacturing, 1*, 77–85.
11. Yampolskiy, M., Schutzle, L., Vaidya, U., & Yasinsac, A. (2015). Security challenges of additive manufacturing with metals and alloys. In *International Conference on Critical Infrastructure Protection*. New York: Springer.
12. Jensen, J. C., Chang, D. H., & Lee, E. A. (2011). A model-based design methodology for cyber-physical systems. In *Wireless Communications and Mobile Computing Conference (IWCMC)*. New York: IEEE.
13. Kothamasu, R., Huang, S. H., & VerDuin, W. H. (2009). System health monitoring and prognostics—A review of current paradigms and practices. In *Handbook of Maintenance Management and Engineering* (pp. 337–362). New York: Springer.
14. Groover, M. P. (2007). *Automation, production systems, and computer-integrated manufacturing*. Upper Saddle River: Prentice Hall Press.
15. Faruque, A., Abdullah, M., Chhetri, S. R., Canedo, A., & Wan, J. (2016). Acoustic side-channel attacks on additive manufacturing systems. In *ACM*.
16. Debar, H., Dacier, M., & Wespi, A. (1999). Towards a taxonomy of intrusion-detection systems. *Computer Networks, 31*(8), 805–822.
17. Cardenas, A., et al. (2009). Challenges for securing cyber physical systems. In *Workshop on future directions in cyber-physical systems security*.
18. Chhetri, S. R., Canedo, A., & Al Faruque, M. A. (2016). KCAD: Kinetic cyber-attack detection method for cyber-physical additive manufacturing systems. In *Proceedings of the 35th International Conference on Computer-Aided Design*. New York: ACM.
19. Evans, B. (2012). *Practical 3D printers: The science and art of 3D printing*. New York: Apress.
20. Hendershot, J. R., & Kuo, B. C. (1993). Causes and sources of audible noise in electrical motors. In *Incremental Motion Control Systems and Devices Symposium*.
21. Yang, S. J. (1981). *Low-noise electrical motors* (Vol. 13). Oxford: Oxford University Press.
22. Timár-P, L. T. P., & Tímár, P. L. (1989). *Noise and vibration of electrical machines* (Vol. 34). North Holland.
23. Gieras, J. F., et al. (2005). *Noise of polyphase electric motors*. Boca Raton: CRC Press.
24. *k-quad 5.1 250 mm quadcopter frame*. (2014). thingiverse. http://www.thingiverse.com/thing:397036

Chapter 6
Data-Driven Security Analysis Using Generative Adversarial Networks

6.1 Introduction

In this chapter, we will present a generative adversarial networks based modeling approach for performing security analysis in cyber-physical systems. The fourth industrial revolution will encompass a large scale use of cyber-physical systems. This manufacturing system in combination with CPS is also known as cyber-physical production systems (CPPS) [1]. CPPS is an integration of sub-systems from multiple cyber and physical domains interconnected through communication networks. Using CPPS will aid in making the factory units smarter and adaptive; however, due to the tight interactions between the cyber and physical domains, there may exist various cross-domain vulnerabilities in CPPS.

There are two types of exploits of cross-domain vulnerabilities that we are primarily interested in: *side-channel attacks* (Chap. 1) and *kinetic cyber-attacks* (Chap. 5). Both types of attacks may even be composed of smaller exploits that target confidentiality, integrity, or availability. Existing CPPS modeling tools were created for the purpose of analyzing the system-level performance, reliability, energy efficiency, and quality of controls. Security is ignored by these tools and left as an afterthought of system design. However, most of the existing CPS security research work only target known vulnerabilities in specific domains. Afterward, they suggest an ad-hoc repair such as patching software and/or replacing hardware components without showing that the repaired system is free of further vulnerabilities [2, 3]. Some of these CPS also end up being a part of the overall CPPS.

Since these solutions cannot analyze and explore all the potential vulnerabilities, they cannot answer the questions such as "How secure is this CPPS against a confidentiality attack by a specific attacker?" or "Can we detect an integrity or availability attack on a CPPS?" To address these questions and improve the CPPS security research, we hypothesize that it is necessary to have a new modeling

© Springer Nature Switzerland AG 2020
S. R. Chhetri, M. A. Al Faruque, *Data-Driven Modeling of Cyber-Physical Systems Using Side-Channel Analysis*,
https://doi.org/10.1007/978-3-030-37962-9_6

approach that takes into consideration the signal and energy flows of a system. We propose a security model that abstracts the relationship between the energy flows and signals flows to answer questions like the aforementioned ones.

6.1.1 Research Challenges

Notice that since the attack models mostly describe the capability of attackers rather than the system, one might assume that the existing attack models may be directly applied in CPPS security analysis. However, there exist the following research challenges to create a system-level model for CPS security analysis:

1. The existing security properties are mostly only proposed for analyzing attacks in the cyber-domain [2]. Thus, analyzing the cross-domain attacks in CPPS requires new types of security properties applicable in both the cyber and physical domains.
2. The existing system-level behavior models in CPPS require various models of computation (MoCs),[1] for the cyber-domain and physical domain (an MoC cannot be cross-domain) [4]. In order to analyze cross-domain security, a unified system behavior of interest for CPPS is required.
3. In the CPPS environment, there are multiple sub-systems interacting with each other; therefore, information leakage or attack detection needs to be performed across multiple sub-systems.

To address these challenges, researchers have proposed to use information flow as the basis of a unified behavior model of cyber and physical domains for security analysis in [5]. However, [5] only focused on the application related to commodity flows and proposed a coarse-grained information flow that models the commodity flow changes as an event. It fails to provide the capability of a quantified analysis of cross-domain attack and detection capability.

Modeling information flows requires a statistical method. However, there is no guarantee that collected data during both design-time and run-time is sufficient to create an accurate security model. In this chapter, we use generative adversarial networks (see Sect. 6.1.2) to acquire a better estimate of the distribution (conditional in this paper) of the data. On one hand, if there is a large amount of data available, the discriminator of the generative adversarial network is able to estimate the data distribution, on the other, the generator, since it never sees the real data estimates the distribution without over-fitting on the limited data, thus providing better distribution estimation. For security analysis, we consider abstracting a CPPS by the signal and energy flows and deriving the conditional densities among flow pairs using conditional generative adversarial network (CGAN).

---

[1]An MoC is a set of allowable operations used in computation and their respective costs (e.g., timing, performance, and memory overhead).

6.1.2 Preliminaries

First, let us present some preliminary definitions about the energy and signal flows as well as GANs before explaining the details about our proposed CGAN-based security model of the CPPS.

Signal Flow: A signal flow is modeled as a discrete signal. It may be defined as a random variable F_S which have n number of possible values, $F_S \in \{f_1, f_2, \dots, f_n\}$. We define a set of n events $E = \{E_1, E_2, \dots, E_n\}$, where $E_i = 1$ when $F_S = f_i$, and $E_i = 0$ otherwise. We assume the probabilities for E_i are known as $Pr(E_i)$.

Energy Flow: We define energy flow as a continuous-domain time-dependent variable F_E. Given a feature construction function $f_X(\cdot)$, we may construct a set of feature vectors $X = f_X(F_E)$. Next, given a feature extraction and selection function $f_Y(\cdot)$, we may extract a set of more relevant feature vectors $Y = f_Y(X)$. Assume $Y = \{Y^1, Y^2, \dots Y^m\}$, where each feature vector Y^i may have n_i number of possible values $Y^i \in \{y_1^i, \dots, y_{n_i}^i\}$. We define a set of n_i events $E_i = \{E_1, E_2, \dots, E_{n_i}\}$, where any event E_{i_j} is true if $Y^i = y_j^i$ is met. We assume the probability for E_{i_j} is known as $Pr(E_{i_j})$.

Generative Adversarial Network: A generative adversarial network (GAN) consists of a generative model G that captures the probability distribution of the flow's data $Pr(F_1)$, and a discriminator model D that estimates the probability that a sample of data x came from the training data rather than the generative model, $Pr(x \in F_1)$ [6]. GAN is a form of a maximum likelihood estimator that is trained to estimate the true probability distribution of a random variable by utilizing the empirical distribution derived from the training data. It achieves this by minimizing the *Kullback–Leibler divergence* between the empirical distribution and the generator's distribution, as follows [6]:

$$\theta^* = \arg \min_{\theta} D_{KL}(Pr_{data}(F_1) \parallel Pr_G(F_1; \theta)) \tag{6.1}$$

where θ are the model parameters. The models used in these networks are generally deep convolutional neural networks. Some advantages of using GAN over other methods are that it does not require Markov chains (where the chance of state explosion is high) and they can generate samples in parallel [6]. A conditional generative adversarial network (CGAN) is a variation of the GAN, where the generator and the discriminator both receive a conditional variable F_2 (see Fig. 6.2). The primary objective of the generative model G is to learn the conditional probability distribution $Pr(F_1|F_2)$. In order to do this, it will require the labeled data $(f_{1_1}, f_{2_1}), (f_{1_2}, f_{2_2}), \dots (f_{1_n}, f_{2_n})$. Here $\{f_{1_1}, f_{1_2}, \dots, f_{1_n}\}, \{f_{2_1}, f_{2_2}, \dots, f_{2_n}\}$ are the values taken by the random variable F_1 and F_2, respectively. G and D have a common objective function based on the two-player mini-max game, as follows [7]:

$$\min_{G} \max_{D} V(D, G) = \mathbb{E}_{F_1 \backsim Pr_{data}(F_1)}[\log(D(F_1|F_2))]$$

$$+\mathbb{E}_{Z \backsim Pr_Z(Z)}[\log(1 - D(G(Z|F_2)))] \qquad (6.2)$$

where Z is a noise random variable that is provided along with F_1 and F_2 to G for training.

In Eq. 6.2, the model G is trained to minimize $\log(1 - D(G(Z|F_2)))$ so that it may generate samples more similar to the training data. D is trained to maximize $\log(D(F_1|F_2))$ so that it may accurately differentiate between the real data and the generated data. Given enough time and capacity, the end result will be that G is trained to accurately estimate the $Pr(F_1|F_2)$ and D cannot differentiate G's output from real or generated data. This is an important concept for our security model because using $Pr(F_1|F_2)$, one can estimate a (*signal/energy*) flow based on another flow.

6.1.3   Novel Contributions

To resolve the challenges of CPPS security modeling, our novel contributions are as follows:

1. **Conditional Generative Adversarial Network-based Model for Security Analysis of CPPS (Sect. 6.2)** which models the conditional probability distribution between the various information flows (cyber-domain data and physical domain energy flow).
2. **Automatic Model Generation Algorithm (Sect. 6.3)** which includes a graph search and pruning algorithm to reduce the complexity of the model and simulation/experimental methods to estimate and generate the proposed behavior model.
3. **A Case Study (Sect. 6.4)** for analyzing information leakage from multiple physical emissions in a single sub-system (additive manufacturing) to demonstrate the applicability of the proposed modeling technique in security analysis.

6.2   CGAN-Based CPPS Security Model

A typical CPPS model (see Fig. 6.1) includes multiple sub-systems $(1, 2, \ldots, n)$. In each sub-system, there are cyber and physical domain components with energy and signal flows connecting them. Moreover, the energy and signal flows may occur among sub-systems as well. In this section, we propose to abstract the relationship between the various flows (*energy–energy*, *signal–energy*, and *signal–signal*), either in a single sub-system or across various sub-systems. This will be achieved by using the conditional generative adversarial model (CGAN). We propose to model

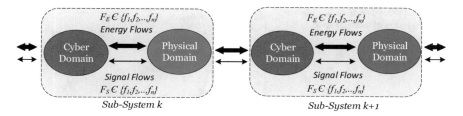

Fig. 6.1 Cyber-physical production system with multiple sub-systems

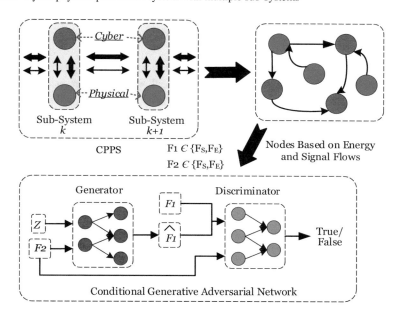

Fig. 6.2 CGAN-based modeling of CPPS

the system behavior using the relationship between the various flows as shown in Fig. 6.2. Based on the particular system design of CPPS, we can acquire all the signal and energy flows during design-time. One can then construct a graph where the various nodes represent the cyber and physical domain components, and the edges represent the signal and energy flow between them (see Fig. 6.3).

The graph created in Fig. 6.3 is then used to list all the flows $F = \{F_1, F_2, \ldots, F_n\}$ with $F \in \{F_S, F_E\}$ that are available in the CPPS system. From this, we extract the flow pairs (F_i, F_j). Each pair is then supplied to the CGAN to model $Pr(F_i|F_j = f_j)$ or $Pr(F_j|F_i = f_i)$. We can infer that a high value of conditional probability indicates a strong relationship between two flows (given knowledge of one of the flows). Based on these distributions, we can therefore analyze the relationship between various flows from a security perspective. For

Fig. 6.3 Decomposition of
CPPS in terms of components
and flows

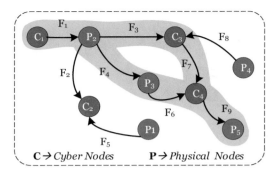

example, using $Pr(F_1|F_9)$ (see Fig. 6.3), one can answer questions similar to the
following: Is data in F_1 (cyber-domain) being leaked from F_9 (physical domain)?
Can F_9 be used to monitor any attacks in the integrity of the flow path from node C_1
to P_5? Various other metrics may also be created using the conditional probability
values (e.g., mutual information metrics of side-channel attacks).

In summary, the proposed CGAN-based security model provides a theoretical
foundation to enable a system-level methodology for the design and analysis of
CPPS against cross-domain attacks. To the best of our knowledge, this is the first
effort towards building a security analysis methodology for CPPS based on CGAN.

6.3   CGAN Model Generation

To address the need for a design-time security analysis tool for CPPS, we propose
to use a two-step method which generates the proposed CGAN-based security
model from a given CPPS model, as shown in Fig. 6.4. The two steps include
(1) graph generation and (2) CGAN-based security model generation. In order to
develop the graph from the existing CPPS, the graph generation algorithm takes the
design-time CPPS architecture as input with information about all the sub-systems
$\{Sub_1, Sub_2, \ldots, Sub_n\}$, cyber C and physical P domain components with each
sub-system, and the corresponding energy flow F_E and signal flow F_S data among
each sub-system.

Step 1: CPPS Graph Generation Based on Signal and Energy Flows First,
it is necessary to have a list of all the signal and energy flows between cyber
and physical components and also between sub-systems. The design-time CPPS
architecture specifications will consist of this information and we will convert it
into a graph-based model. A graph would allow us to select the energy and signal
flow pairs necessary to create the CGAN model in the next step. We will denote the
corresponding graph generated in this step as G_{CPPS} consisting of nodes made of
cyber and physical components.

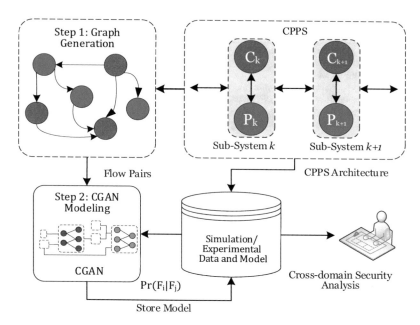

Fig. 6.4 Automatic model generation method

Building the graph, G_{CPPS}, given the design-time architecture of the CPPS, is straightforward and presented in Algorithm 1. After building G_{CPPS}, in Lines 12–19 we reduce the graph to find only relevant pairs that can be modeled using CGAN.

Step 2: CGAN Model Generation We assume that the given G_{CPPS} is a black box with only input/output flows. There may exist historical data of the CPPS that we can directly use (e.g., testing/run-time data of a system). The $Pr(F_1|F_2)$ learned by the CGAN relies on the training data. In CPPS, this training data is bound by either the architecture or the production capability. Hence, during design-time security analysis, a large amount of training data is required. This may be a drawback of the CGAN-based modeling, but this problem persists in other data collection methods as well. In CPPS, there are some limitations to the access of the high-level data which needs to be protected, such as product specification, due to the mechanical structure of the sub-systems. This allows us to gather training data within the specified bound and estimate $Pr(F_1|F_2)$ using our model. The amount of data given for training can also be modified according to the attacker capability or attacks detection models resources, such as sub-system type, money, or time. Algorithm 2 is used to train the CGAN. The algorithm takes the flow pair list FP_T generated in Algorithm 1 and the data corresponding to the flows in the CPPS.

Algorithm 1: CPPS graph and flow pairs generation

    Input: CPPS Architecture Data: Sub, C, FP, F_S, F_E
    Input: Historical Data: $Data$
    Output: Flow Pair List: FP_T
 1 Initialize G_{CPPS} with V nodes and its adjacency list, E
 2 Initialize FP_F as flow pairs (F_i, F_j) list
 3 Initialize FP_T as flow pairs (F_i, F_j) list
 4 Remove all the feedback loops so that signal and energy flows are directed
 5 **foreach** *Subsystem $S \in Sub$* **do**
 6   Node list Q=(Add all $P_i \in S$ and $C_i \in S$)

 7 **foreach** *Node $v \in Q$* **do**
 8   Add v in G_{CPPS}
 9   **foreach** *Node $u \in Q \setminus v$* **do**
10   **if** *F_S or F_E exists between v and u* **then**
11   Add u into adjacency list of v, $E[v]$

12 **foreach** *Flow $F_{1_i} \in E$* **do**
13   **foreach** *Flow $F_{2_j} \in (E \setminus F_{1_i})$* **do**
14   **if** *head of F_{2_j} is reachable from tail of F_{1_i} using DFS* **then**
15            $FP_{i,j} = (F_1, F_2)$
16            $FP_F = FP_F \cup FP_{i,j}$

17 **foreach** *$FP_{i,j} = (F_{1_i}, F_{2_j}) \in FP_F$* **do**
18   **if** *$FP_{i,j} \in Data$* **then**
19         $FP_T = FP_T \cup (FP_{i,j})$

20 **return** FP_T

For each of the flow pairs available in FP_T, Lines 1–11, which are based on [6, 7], iteratively take sample flows from F_i and F_j, and train D by stochastic gradient ascent (Line 8), and train G by stochastic gradient descent (Line 10). The number of steps and the iterations to be performed depends on the assumptions about the attacker and can be easily modified accordingly. At the end, G learned for each flow pair is returned and stored. This model effectively will learn the conditional probability $Pr(F_i|F_j)$ [6, 7].

The convergence of G to the historical data has been extensively proven in [6, 7]. Then based on the assumption of how much actual data an attacker can acquire, or how much data an attack detection model can acquire, $Pr(F_i|F_j)$ can be estimated accordingly by adjusting D in the training phase.

6.4   Case Study and Analysis

As a case study, we provide the CGAN-based modeling for security analysis of a sub-system of a CPPS. The CPPS sub-system selected is a fused modeling deposition-based additive manufacturing system, also known as a 3D printer.

Algorithm 2: CGAN model generation and storage for security analysis

---

Input: Flow Pairs: FP_T
Input: Historical Data: $Data$
Input: CGAN Training Parameters: Batch Size n, Step Size k, Number of Iterations $Iter$
Output: Generator, Discriminator: G, D

1 **foreach** $FP_j \in FP_T$ **do**
2   Extract flows corresponding to (F_1, F_2) of $FP_j \in Data$
3   **for** $Iter$ steps **do**
4   **for** k steps **do**
5   Acquire n mini-batch noise samples $\{z_1, z_2, \ldots, z_n\}$ from $Pr(Z)$
6   Acquire n mini-batch noise samples $\{f_{1_1}, f_{1_2}, \ldots, f_{1_n}\}$ from $Pr_{data}(F_1)$
7   Corresponding to $\{f_{1_1}, f_{1_2}, \ldots, f_{1_n}\}$, acquire $\{f_{2_1}, f_{2_2}, \ldots, f_{2_n}\}$ from $Pr_{data}(F_2)$
8   Update D, by ascending its stochastic gradient:
                $\nabla_{\theta_d} \frac{1}{n} \sum_{i=1}^{n} [\log D(f_{1_i}|f_{2_i}) + \log(1 - D(G(z_i|f_{2_i})))]$
9   Acquire n mini-batch noise samples $\{z_1, z_2, \ldots, z_n\}$ from $Pr(Z)$ and the $\{f_{2_1}, f_{2_2}, \ldots, f_{2_n}\}$ used in the Line 17
10   Update G, by descending its stochastic gradient:
11         $\nabla_{\theta_g} \frac{1}{n} \sum_{i=1}^{n} [\log(1 - D(G(z_i|f_{2_i})))]$

12 **return** G that estimates $Pr(F_i|F_j)$, and D

Additive manufacturing has been predicted to be one of the enabling technologies for the next generation of CPPS due to its rapid prototyping and distributed manufacturing capability. However, many emerging threats to this CPPS sub-system have also been highlighted [8, 9]. Hence, we will demonstrate how some security analysis may be performed in this sub-system using the proposed CGAN-based modeling.

The state-of-the-art Cartesian 3D printer, as shown in Fig. 6.5, consists of four stepper motors. Three of them are used to provide printer nozzle movement in the X, Y, and Z directions, while the fourth motor is used to extrude the filament while printing. The speed and direction of all the stepper motors are controlled by cyber-domain instructions written with G-code, a programming language widely used in industrial systems to tell a tool what to do and how to do it, along with M-code, auxiliary commands for miscellaneous machine functions. Our experimental setup is enclosed in a makeshift anechoic chamber to isolate the noise from the environment and other components in the lab. A contact microphone (C411 L) is attached to the back of the 3D printer to acquire the acoustic energy flow dissipated to the environment.

6.4.1 G_{CPPS} Generation

Using Algorithm 1, the additive manufacturing CPPS can be decomposed to its corresponding energy and signal flows in the form of a graph, G_{CPPS} (see Fig. 6.6).

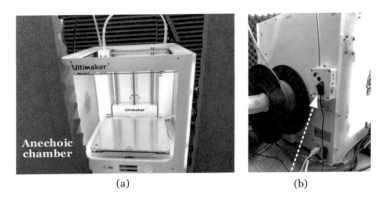

(a)   (b)

Fig. 6.5 Additive manufacturing as a sub-system for security analysis. (**a**) Ultimaker 3D printer.
(**b**) Acoustic sensor

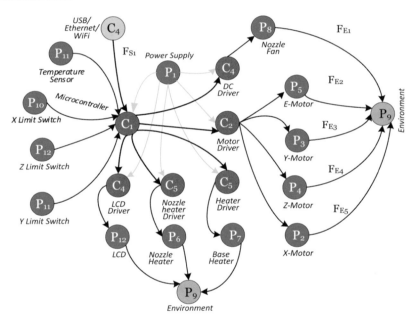

Fig. 6.6 Graph generation for cyber-physical additive manufacturing system

One may notice that vertices C_4 and P_9 are not components of the 3D printer. C_4 represents the external signal flows from other sub-systems into the 3D printer. The node P_9, on the other hand, represents the physical environment. Various energy flows that are either intentional or unintentional passing to the environment are encompassed by the edges going towards the node P_9. From this graph, flow pairs P_T is extracted using Algorithm 1.

6.4.2 Experimental Data Collection

In our experiment, we have only selected cross-domain flow pairs for security analysis. Specifically, we analyze the acoustic/vibration energy in the physical domain and the G/M-code instructions passed to the 3D printer sub-system from an external node. The G/M-code instructions are the signal flows entering the sub-system from node C_4. In our analysis, we have monitored the energy flows between nodes P_2, P_3, P_4, P_5, P_8 and the node P_9. For ease of training the CGAN, we convert the time-domain acoustic energy flows values into frequency domain values using continuous wavelet transforms, which preserves the high-frequency resolution in time domain as well. We obtain a non-uniformly distributed 100 bins $Freq = [freq_1, freq_2, \ldots, freq_{100}]$ between 50 and 5000 Hz (this range may be changed for further security analysis purposes). In this case-study analysis, for simplicity, we extract G/M-codes from 3D objects that only move one stepper motor at a time. The G/M code is one-hot encoded based on presence of instructions that run stepper motors X ([1, 0, 0]), Y([0, 1, 0]), and Z([0, 0, 1]), respectively. This encoding is done based on G/M-codes G_t and G_{t-1}. For example, if G_{t-1} is $[G_1 \; F1200 \; X5 \; Y5 \; Z5]$ and G_t is $[G_1 \; F1200 \; X10 \; Y5 \; Z5]$ then encoding for G_t will be $Cond = [1, 0, 0]$ as only stepper motor X will run. For more thorough security analysis, the one-hot encoding can be extended to consider a combination of signal and energy flows. For example, for three physical components and their combination, the one-hot encoding can be of size $2^3 = 8$.

6.4.3 CGAN Modeling

Using the one-hot encoding obtained in Sect. 6.4.2 as conditions $Cond$, we have trained CGAN to estimate $Pr(Freq|Cond)$. In doing so, we can derive a trained conditional density function to find the probability of a particular value of frequency component $freq$ given which stepper motor is running. These conditions are generated from the signal flows coming from node C_4, and frequencies extracted from the acoustic energy flows going to the environment node P_9 from the various nodes. Hence, the conditional density function estimates the relation between the signal flow from node C_4 to C_1 and energy flows from nodes P_2, P_3, P_4, P_5, P_8 to the node P_9.

Figure 6.7 shows the training results for the CGAN. On the X axis, the iteration number is increasing. With the increasing iteration, however, the more signal and energy pair data are also incorporated. We can observe that initially, G's loss is high, whereas D's loss is low. However, over more iterations and data, the G's loss decreases, making it difficult for D to know whether the data generated is real or fake, and hence increasing the loss of D.

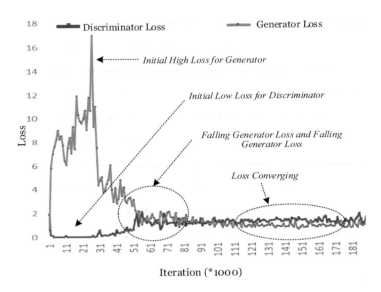

Fig. 6.7 Training results for the CGAN

6.4.4   Security Analysis Results

In our experiment we will demonstrate how the CGAN modeling may be used for security analysis concerned with *confidentiality* breach through the side-channels, and design of *integrity* and *availability* attack detection on the physical components of the 3D printer sub-system using the physical domain (or the same side-channels). Both of these analyses have one thing common, the conditional relationship between the energy flows given the signal flows. In this section, we will show how the CGAN model (discriminator D, generator G, and noise Z) can be used to check this conditional relation for security analysis. The preliminary algorithm used for security analysis shown in Algorithm 3 (this may be changed for more complex signal flow analysis but can still use the same CGAN).

Algorithm 3 computes security metrics based on the average likelihood of test samples. If a test sample x has a high likelihood for a specific conditional distribution $Pr(x|Cond_i)$ and its actual label is $Cond_i$, then it means that there is a strong relationship between $Cond_i$ and the x. This type of metric can be used to evaluate the system's confidentiality vulnerability level or perhaps to detect an integrity or availability attack. For each condition label $Cond_i$, we first generate samples X_G with $G(Z|Cond_i)$. Then for a given Parzen window size h, and current feature index $FtIdx$, we create an estimated conditional distribution $FtDistr = Pr(X_G^{FtIdx}|Cond_i)$ via the Parzen Gaussian Window method (Line 10). For each frequency feature, we derive the corresponding test samples from our test batch X_{test} (Line 8), and for each test sample, we update two metrics based on the likelihood.

Algorithm 3: Algorithm for the security analysis

Input: $D, G, Z, Cond, GSize, X_{test} \in \mathbb{R}^{LxM}$
Input: Frequency Feature Indices: $FtIndices \in \mathbb{R}^{K}$
Input: Parzen Window Width: h
Output: Average Likelihood Metric Sets: $AvgCorLikes, AvgIncLikes$
1   Initialize $AvgCorLikes$ as $\mathbb{R}^{NxK}$ matrix
2   Initialize $AvgIncLikes$ as $\mathbb{R}^{NxK}$ matrix
3   **foreach** $C_i \in Cond$ **do**
4     $CorLike = 0, CorNum = 0$
5     $IncLike = 0, IncNum = 0$
6   **foreach** $FtIdx \in FtIndices$ **do**
7   X_G = generated $GSize$ samples from $G(Z|C_i)$
8   **foreach** $X_{test}^{l,FtIdx}$, where $l \in [1, \ldots, L]$ **do**
9         $FtDistr = ParzenWindowDensity$
10         $(X_G^{FtIdx}, 'Gaussian', h,)$
11         $LogLike = FtDistr.score(X_{test}^{l,FtIdx})$
12         $Like = exp(LogLike) * h$
13   **if** $Label(X_{test}^{l,FtIdx}) == Cond_i$ **then**
14           $CorLike += Like$
15           $CorNum += 1$
16   **else**
17           $IncLike += Like$
18           $IncNum += 1$
19       $CurrAvgCorLikes = CurrAvgCorLikes \cup \{CorLike / CorNum\}$
20       $CurrAvgIncLikes = CurrAvgIncLikes \cup \{IncLike / IncNum\}$
21     $AvgCorLikes[i] = CurrAvgCorLikes$
22     $AvgIncLike[i] = CurrAvgIncLikes$
23   Reset $CurrAvgCorLikes$ and $CurrAvgIncLikes$
24   **return** $AvgCorLikes, AvgIncLikes$

The first metric $CorLike$ is the total likeliness that this test sample belongs to $FtDistr$ and the other $IncorLike$ refers to the total likeliness that it does not belong to $FtDistr$. We then average the two metrics according to the number of test samples per feature (Lines 19–20). In the outermost loop based on the conditions, we update two sets $AvgCorLikes$ and $AvgIncLikes$, with the corresponding sets of averaged metric values created in the inner loops. Higher values in the first set mean that our model has learned a better relationship between data (e.g., features) and their correct conditions (e.g., running motors). However, higher values in the second set mean that our model has learned unexpected and potentially unrealistic relationships between data and other conditions.

The conditional density functions estimated by the CGAN is shown in Fig. 6.8. The frequency magnitudes for the $Freq = [freq_1, freq_2, \ldots, freq_{100}]$ are scaled between 0 and 1. The Parzen window width h value used for the Gaussian kernel density estimation is 0.2. Hence, the actual probability of the frequency values is obtained by multiplying it by 0.2.

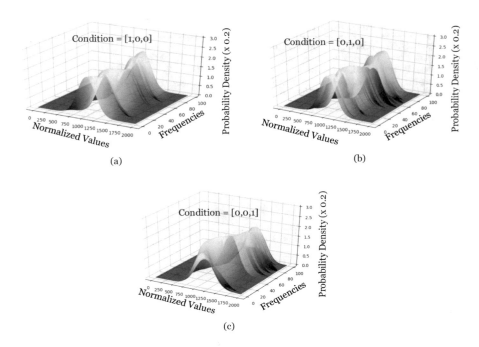

Fig. 6.8 Conditional probability distribution for the acoustic energy signal (h = 0.2). (**a**) Conditional probability for acoustic emissions for Stepper Motor X. (**b**) Conditional probability for acoustic emissions for Stepper Motor Y. (**c**) Conditional probability for acoustic emissions for Stepper Motor Z

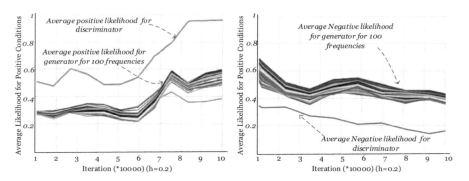

Fig. 6.9 Average correct and incorrect likelihood values for the conditions

In Fig. 6.9, the average correct and incorrect likelihood values are presented for the condition ([1, 0, 0]). As can be seen, over increasing iterations, the positive likelihood averages improve. This shows that the generator is able to accurately learn the conditional distribution of the acoustic emissions according to the signal flows.

Table 6.1 Average and incorrect likelihood of acoustic energy flows

	$h = 0.2$		$h = 0.4$		$h = 0.6$		$h = 0.8$		$h = 1$	
	Cor	Inc	Cor	Inc	Cor	Inc	Cor	Inc	Cor	Inc
Cond1	0.6000	0.2245	0.6000	0.3247	0.6069	0.3634	0.6293	0.3783	0.6437	0.3856
Cond2	0.5750	0.3887	0.5750	0.3961	0.5750	0.3974	0.5750	0.3982	0.5532	0.3978
Cond3	0.6556	0.3876	0.6556	0.3956	0.6556	0.3979	0.6601	0.3983	0.6556	0.3985

Based on the density function estimated using the generator of the CGAN, we have used the security analysis Algorithm 3 to calculated the average correct and incorrect likelihoods of the emissions given the three conditions (presence of X, Y, or Z motor movement in the G/M-code). Using this, a CPPS designer can estimate if an attacker is able to estimate the G/M-code based on acoustic emissions. For example, in Table 6.1, we have presented the average correct and incorrect likelihoods of a single feature in the frequency domain based on the G/M-code related to the X, Y, or stepper motor. The table shows that an attacker can estimate condition 3, which is the presence of the Z-motor movement in the G/M-code, better than the other conditions (presence of X or Y motor movement). Moreover, if a designer needs to create an integrity and availability attack detection model to detect attacks on individual components (X, Y, or Z motor) using the side-channels, he/she will be able to estimate the performance of such a model using the CGAN model.

6.5   Summary

In this chapter, we have presented GAN-Sec, a conditional generative adversarial network (CGAN) based modeling approach for the security analysis of cyber-physical production systems (CPPS). To do this, GAN-Sec abstracts the system in terms of its flows and estimates the conditional distribution between them using a CGAN model. We have used GAN-Sec to analyze the security of an additive manufacturing CPPS sub-system. The promising results indicate that GAN-Sec is applicable in analyzing the cross-domain security of CPPS.

References

1. Monostori, L. (2014). Cyber-physical production systems: Roots, expectations and R&D challenges. In *Procedia CIRP*.
2. Cardenas, A., Amin, S., Sinopoli, B., Giani, A., Perrig, A., & Sastry, S. (2009). Challenges for securing cyber physical systems. In *Workshop on future directions in cyber-physical systems security*.
3. Washington   Post.   http://www.washingtonpost.com/wp-dyn/content/article/2008/06/05/AR2008060501958.html

4. Sztipanovits, J., Bapty, T., Neema, S., Howard, L., & Jackson, E. (2014). Openmeta: A model- and component-based design tool chain for cyber-physical systems. In *From Programs to Systems. The Systems perspective in Computing* (pp. 235–248). New York: Springer.
5. Akella, R., Tang, H., & McMillin, B. M. (2010). Analysis of information flow security in cyber– physical systems. *International Journal of Critical Infrastructure Protection, 3*(3), 157–173.
6. Goodfellow, I., Pouget-Abadie, J., Mirza, M., Xu, B., Warde-Farley, D., Ozair, S., et al. (2014). Generative adversarial nets. In *Advances in Neural Information Processing Systems.*
7. Mirza, M., & Osindero, S. (2014). Conditional generative adversarial nets. Preprint. arXiv:1411.1784.
8. Sturm, L., Williams, C., Camelio, J., White, J., & Parker, R. (2014). *Cyber-physical vulnerabilities in additive manufacturing systems.* Context.
9. Zeltmann, S. E., Gupta, N., Tsoutsos, N. G., Maniatakos, M., Rajendran, J., & Karri, R. (2016). Manufacturing and security challenges in 3D printing. *JOM,* 1–10. https://doi.org/10.1007/ s11837-016-1937-7

Part III
Data-Driven Digital Twin Modeling

Chapter 7
Dynamic Data-Driven Digital Twin Modeling

7.1 Introduction

Cyber-physical manufacturing systems have various cyber processes interacting with the physical domain components through a communication network [1]. The processes in the cyber-domain can be monitored for each clock cycle of the computing and the communicating component. This allows us to accurately predict the behavior of the cyber-domain components. However, in the physical domain, the same is not true. Monitoring physical domain components and predicting the physical system behavior are faced with a myriad of challenges [2, 3]. One of the major challenges arises due to the fact that the physical domain continuously interacts with the environment, humans, etc., which changes the states of the physical components [4]. Predicting this interaction and the effects of the environment has been one of the major research topics in the manufacturing industry [5–8]. Being able to infer about this interaction is crucial as it affects the process parameters which eventually determine the quality of the products.

To overcome the problem of process control, among many others, digitization of the physical domain can be one of the solutions [2, 9–11]. This real-time digital representation of the physical domain, also known as "The Digital Twin" [3, 12, 13], will make it is easier for manufacturers to accurately predict the future system performance and plan accordingly as well [2, 14, 15]. The future trend will be in using the capability of the Internet-of-Things in producing a massive amount of data to create a digital twin that interacts with the cyber-domain [16, 17]. The concept of digital twin was first used by NASA, whereby they wanted a digital twin replica of the physical system used for the space exploration [18]. Since then, various works have been done to create the digital twin of the physical twin [12, 19]. Authors in [20] have created digital twin, an ultrahigh fidelity model of individual aircraft, to predict the structural integrity and the life of the aircraft structure. A company like IBM is providing sensor systems for creating the digital twin of buildings [21]. In

© Springer Nature Switzerland AG 2020
S. R. Chhetri, M. A. Al Faruque, *Data-Driven Modeling
of Cyber-Physical Systems Using Side-Channel Analysis*,
https://doi.org/10.1007/978-3-030-37962-9_7

fact, technology solutions are hitting the market for enabling the creation of digital twin. Ansys [22] released their technology solution for building the twin, with an example for creating the digital twin for pumps. Honeywell has recently presented a connected plant concept with a solution to bring all process domain knowledge to create a digital twin in the cloud [23]. Digital twin technology is also being used for life cycle engineering asset management in [24]. Authors in [25, 26] provide analysis for research needs and current status for the building blocks of the first generation of the digital twin for additive manufacturing (or a 3D printer) system.

7.1.1   Research Challenges

In summary, the benefits of building and updating a digital twin of a physical system have recently been acknowledged both by researchers and industries. However, the digital twin technology is still in its infancy, and it faces various research challenges such as:

- Understanding what variables in the physical domain can be extracted.
- Knowing how to select the number and position of sensors either during design-time or during the usage time (for legacy systems).
- Understanding how a digital twin model can be developed given the constraints of resources (such as sensors).
- Figuring out when to update the digital twins (as lightweight as possible to meet the resource constraints), to make sure that they can accurately predict the system performance.

7.1.2   Contributions

In order to solve these research challenges, in this chapter, we provide the following various solutions:

- Present analysis of physical domain signals that can be extracted from cyber-physical manufacturing systems (Sect. 7.5.1.1).
- Perform feature engineering on signals and data-driven modeling for creating the digital twin (Sect. 7.4.2).
- Use dynamic data-driven application systems for keeping the digital twin up-to-date and lightweight (by re-ranking and selecting the most prominent features for building the digital twin) (Sect. 7.4).

The major contribution of this chapter is in providing a methodology for keeping the digital twin alive using the concept of dynamic data-driven application systems (DDDAS) [27]. The DDDAS concept is used in influencing the data-driven models (in our case the digital twins), by providing dynamic feedback in updating the models based on the real-time data from the physical domain. The key technical challenge of modeling and updating a living digital twin of a physical twin is solved with the use of dynamically steered sensor data processing (selection and fusion), which uses concepts of DDDAS.

7.1.3   Digital Twin Model

The major contribution presented in this chapter is the creation of the digital twin based on the dynamic feature selection using side-channel emissions. Before building the digital twin, we need to set up an expected outcome or a use case for the digital twin. An optimistic version of the digital twin would be able to demonstrate and predict every possible physical state of the system. However, to narrow down the scope, we set two objectives for the digital twin:

- Based on the status of the physical components, predict the KPIs such as surface texture and dimension of the 3D object that will be printed in the future.
- Be able to know when the digital representation of the physical components is not up-to-date for the KPI predictions.

7.2   Digital Twin of Cyber-Physical Additive Manufacturing System

With these narrowed down scopes, let us explain how the digital twin can be modeled in a cyber-physical additive manufacturing systems. In Fig. 7.1, we present the digital twin representation. It takes into consideration the various environmental interactions and aging phenomena (represented by random variables $B_1, B_2, \ldots, B_n$), the process (random variable α) and design parameters (random variable β), the historical and current analog emissions from the systems (represented by random variable: Acoustics (A_1), Vibration (A_2), Power (A_3), Magnetic (A_4), etc.), and predict the KPI, such as surface texture (represented by a random variable K_1) and dimension (represented by a random variable K_2).

In doing so, it explains the effect of environmental parameters ($B_1, B_2, \ldots, B_n$) on process and design parameters (α, β) which in return affects the KPIs. With this backdrop, the digital twin can be modeled using a machine learning algorithm that explains the relation between analog emissions, process parameters, environmental factors, and the KPIs.

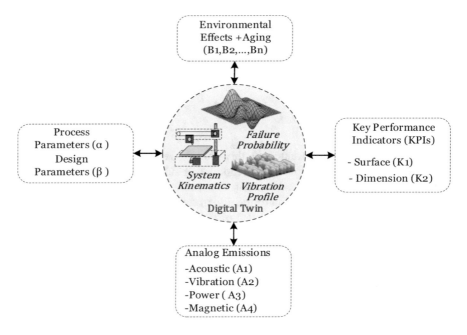

Fig. 7.1 Digital twin of cyber-physical additive manufacturing system

7.2.1 Key Performance Indicators (KPIs)

The two KPIs used in measuring the performance of the additive manufacturing systems are as follows:

7.2.1.1 Surface Texture (K_1)

The quality of the surface texture of the printed object is one of the most sensitive performance indicators that are affected by different design and process parameters. For instance, a slightly higher temperature of the base plate or extruder can cause distortion on the surface by affecting the property of the filament being deposited. Moreover, the slice thickness of the 3D object and the road width of the segment being printed on the XY-plane also cause the surface texture to vary in quality. For quantifying the surface quality, we explored an innovative metric called *dispersion of directionality* (K_1). It is a heuristic metric calculated by performing various image processing algorithms on the surface image. *We would like to emphasize that other standard metric can be used with our methodology (as the methodology is independent of the metric used to measure the KPI), and in fact may improve the accuracy of the digital twin.* The steps to calculate K_1 are as follows:

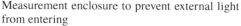

Measurement enclosure to prevent external light
from entering

Internal view of the enclosure

Fig. 7.2 Constant lighting environment for measuring the quality of surface texture

1. **Create a constant lighting environment:** As shown in Fig. 7.2, we use an
   empty box with a small opening for fitting a DSLR camera on the top (this
   camera can be replaced with a low-cost camera as well). This enclosure prevents
   the time-varying external light from entering and affecting the surface texture
   measurement. To maintain constant lighting, we used two similar light sources,
   around 10 *lumen* of luminous flux, placed directly opposite to each other to
   provide homogeneous lighting on the surface of the 3D object. Moreover,
   guidelines are drawn to always place the 3D object on the same location for
   the surface texture measurement.
2. **Remove the background:** Since the 3D objects have a distinct color (green)
   compared to the background (brown), we first transform the image taken by
   the camera from the RGB to the lab color space [28]. Lab color space consists
   of three dimensions: L for the lightness of the image, and a and b for the
   color opponents green-red and blue-yellow, respectively. After this transform,
   we choose the value of a which eliminates the shadows and the brown colored
   background. Then, we perform a constant threshold to create a mask, which
   matches the green colored object. This mask is then applied to the image to
   eliminate the background.
3. **Surface division:** The image is divided into either 4 by 4 or 16 by 16 equal
   parts for aiding the process of mapping the surface texture K_1 value to its
   corresponding analog emissions.
4. **2D discrete Fourier transform:** The discrete transform ($F[k, l]$) is calculated
   for the image of the 3D object's surface ($f[m, n]$) (see Fig. 7.3b), using Eq. 7.1,
   where M and N are the height and the width of the image calculated in the step
   3. The maximum values for k and l are M and N as well.

$$F[k, l] = \frac{1}{MN} \sum_{m=0}^{M-1} f[m, n]e^{-j2\pi(\frac{k}{M}m+\frac{l}{N}n)} \tag{7.1}$$

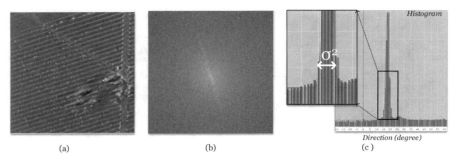

(a)   (b)   (c)

Fig. 7.3 Last three steps (left to right) for measuring the quality of surface texture. (**a**) Surface Texture Image. (**b**) Discrete Fourier transform of the image. (**c**) Distribution parameter estimation

5. **Directionality histogram calculation:** Based on the value of $F[k, l]$, calculated in the previous step, a directionality histogram is calculated using the approach mentioned in [29] for faster calculations (see second image in Fig. 7.3).
6. **Fit normal distribution:** A normal curve is then fitted to the histogram (see Fig. 7.3c) obtained in the previous step, and the corresponding distribution parameters are calculated. Out of these parameters, standard deviation σ is the dispersion metric we use to measure the directionality and hence the surface texture K_1.

The heuristic dispersion metric used for surface texture measurement K_1 does not have a linear relation with a number of the lines in the image or the amount of disorientation in the image. In order to prove the effectiveness of the dispersion metric for surface texture measurement, we asked two people to sort 28 surface images of the 3D objects printed by the fused deposition modeling based additive manufacturing system. These surface textures were varied randomly with various level of surface quality. This sorting was then compared to the sorting order created using the values of *dispersion of directionality* calculated using our algorithm. The result was a similar sorting order. However, due to the subjective nature of quality perception, further work is needed to compare the result of the standard metric and the *dispersion of directionality* for surface texture measurement. Nonetheless, in this chapter, *dispersion of directionality* is used as a KPI for surface texture measurement, due to its ability to represent the varying road thickness of the filament deposited on the XY-plane, presence of non-directional surface pattern, etc. Moreover, the use of standard metric will only make the digital twin modeled using our approach more accurate.

7.2.1.2 Dimension (K_2)

The dimensional accuracy is affected by various process and design parameters. The process parameters that affect K_2 are *build environment temperature, filament feed-rate*, *nozzle temperature*, etc. Whereas, the design parameters that affect the K_2 are *road width*, *slice thickness*, *air gap*, *build orientation*, *raster patterns*, etc. Various environmental factors such as *room temperature*, *humidity*, *vibrations*, etc. can influence these process and design parameters, which in return affect K_2. Since the resolution of the 3D printer used in our experiment is about 12.5, 12.5, and 2.5 μ in X, Y, and Z directions, respectively, micrometer is used to measure the dimension of the 3D objects.

7.3 Keeping Digital Twin Updated

The digital twin model explains the relationship between KPIs, analog emissions, environmental factors, process parameters, and design parameters. In order to keep the digital twin updated, these models will have to be continuously updated based on the run-time difference (δ) between the predicted KPIs and the measured KPIs (as shown in Fig. 7.4). This δ will give the digital twin an intuition of liveliness of the digital twin and enable it to have the cognitive ability. If the error between the predicted and the real KPI value is large then the digital twin will be assumed to be outdated. This in return will trigger feedback to the dynamic feature selection algorithm to re-rank and fuse the feature values to measure the relation between the most recent texture value and the analog emission values calculated, while only selecting the features that are necessary (thus keeping the digital twin model lightweight). The steps for updating the digital twin (as shown in Fig. 7.4) are as follows:

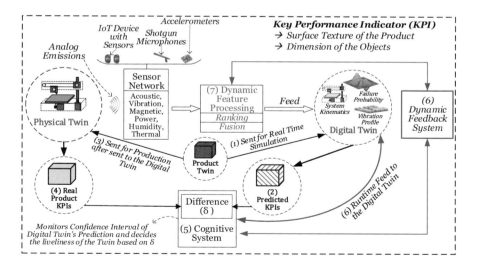

Fig. 7.4 Dynamic data-driven application systems enabled digital twin modeling

1. The product digital twin (3D object) is given to the digital twin of the 3D printer.
2. digital twin of the 3D printer predicts the KPIs (K_1 and K_2).
3. The 3D Object is sent to the physical twin for production only when the predicted KPIs are within a tolerable range.
4. Actual KPIs are measured from the printed 3D Object if it is printed.
5. The cognitive algorithm measures the difference between the actual and predicted KPIs.
6. If the δ is large, feedback is sent to the digital twin and the feature processing algorithms.
7. Dynamic feature processing algorithm re-ranks and fuses the features to get the most up-to-date digital twin model.

The feedback to the data-driven digital twin model is activated by the analog emission data gathered from the sensors. This data represents the current physical state of the system (for example, mechanical degradation may result in variation of the analog emission pattern).

7.4 Building Digital Twin

7.4.1 Sensor/Emission Modality Selection

The 3D printer consists of a set of actuators, mechanical moving parts, heating elements, and a controller board. An actuator under the load consumes energy, vibrates, emits electromagnetic waves, and produces audible sound. The heating elements of the 3D printer, which are embedded inside the nozzle(s) and the base plate, consume electrical energy and convert it into thermal energy. Based on these facts, we decided to monitor the physical domain by acquiring acoustic, electromagnetic, vibration, power, humidity, and temperature data and analyzed them for the amount of information revealed about the cyber-domain and physical domain states of the 3D printer. The validation for the analog emission modality selection is experimentally validated through the accuracy of the digital twin models.

7.4.2 Feature Engineering

After selecting the emissions to be monitored in the physical domain, various features have to be extracted to reduce the size of the raw data collected, and to improve the performance of the machine learning models (or the data-driven models) used to create the digital twin. In this section we will briefly describe the various features extracted in various domains.

7.4.2.1 Time Domain

The various features extracted in time domain are *energy, energy entropy, mean amplitude, maximum amplitude, minimum amplitude, median amplitude, mode of amplitude, peak to peak features* (*highest peaks, peak widths, peak prominence,* etc.), *root mean square values, skewness, standard deviation, zero crossing rate, Kurtosis,* etc. (total 114). These are extracted for each emission. Each of these features capture various properties of the signal, explaining each one of them is out of scope of this chapter, and will be left for the future work.

7.4.2.2 Frequency Domain

Various analog emissions are monitored in the physical domain. Each of them has a different frequency range and characteristics. To capture all of the characteristics for all the signals, various frequency domain signals are analyzed.

- **Frequency Characteristics and Short Term Fourier Transform (STFT):** The frequency characteristics analyzed are based on the short term windows and various characteristics of the frequency domain such as: *mean frequency, median frequency, signal-to-noise ratio, power bandwidth, spectral centroid, spectral entropy, spectral flux, spectral roll off,* etc. (82 in total).
- **Frequency Characteristics and Continuous Wavelet Transform (CWT):** The challenge with short term Fourier Transform based features is the trade-off between time and frequency. The window frame (in the time domain) and the resolution of frequency domain features are highly dependent, and one has to be compromised for the other. Instead of compromising these we have also analyzed continuous wavelet transform in the frequency domain, and analyzed various characteristics of the transform (in Total 58). Based on the result of the digital twin prediction and feature ranking, in the future, the continuous wavelet transform will be used to calculate the discrete wavelet transform with specific approximation and detailed coefficients.

7.4.3 Sensor Positioning

In order for us to determine the position of the sensors selected to acquire the data from the physical domain, we performed profiling of the analog emissions from each modality. For this, a classifier is modeled to classify various cyber-domain data (G/M-codes), such as movement in each axis. Then, various locations around the 3D printer (9 locations for each modality except power, humidity, and temperature) were selected for measuring the analog emission and the corresponding classification scores. A Matlab based graphical user interface is created to analyze the classification scores and the corresponding feature ranking scores for analyzing and finalizing the sensor positions. In Fig. 7.5, the result of the sensor position GUI and the four channel's classification scores are presented. It can be seen from the figure that for channel 1, position 9 gives the highest classification accuracy, whereas for channel 2, channel 3, and channel 4, the position most favorable for higher accuracy are 6, 5, and 2, respectively.

7.4.4   Data-Driven Models

For creating the data-driven models of the digital twin, we explored various machine learning algorithms such as gradient boosted regressor [30], decision tree regressor [31], K nearest neighbor regressor [32], AdaBoost regressor [33], etc. These models were used to model the relationship between the design and process parameters and the analog emissions, and the KPIs.

7.5   Experimental Setup

The experimental setup for modeling and updating the digital twin is shown in Fig. 7.6. As a test case, we have selected fused deposition modeling based cyber-physical additive manufacturing systems. The various components of the experiments are explained as follows:

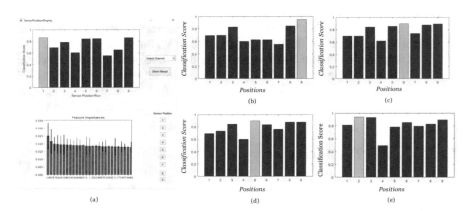

Fig. 7.5 Classification scores for various sensor position. (**a**) Sensor position data GUI in matlab. (**b**) Channel 1. (**c**) Channel 2. (**d**) Channel 3. (**e**) Channel 4

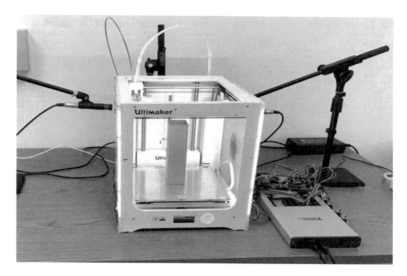

Fig. 7.6 Experimental setup for the digital twin modeling using DDDAS

7.5.1   The Test-Bed

The test-bed consists of an *Ultimaker 3* 3D printer, a set of sensors with analog and digital interface, a data acquisition (DAQ) device, two *Arduino Uno* microcontroller boards coupled with MCP4725 for digital to analog conversion (DAC) purpose, and a personal computer for managing the acquired data. The details of the test-bed are as follows:

7.5.1.1   Sensors

In this chapter, as presented in Table 7.1, we monitor the 3D printer and its surrounding environment using a total of 25 sensors. The maximum sampling rate, sensitivity, and cost are the important factors that determine the type of sensor selected for each modality. We have used three accelerometers for measuring the vibration in the system. These accelerometers have a sampling rate of 1 kHz. A contact microphone is used to measure the acoustic noise and the high-frequency vibration from the printer. Three microphones with a sampling rate of 20 kHz are used to measure the acoustic emissions in the audible range from the 3D printer. ZOOM TAC-8 phantom power supply and amplifier are used to condition the acoustic signal before feeding it to the DAQ. The electromagnetic field intensity variation caused by the stepper motors of the 3D printer cannot be captured without using high precision EM sensors. Instead, we use three compass sensors, which are designed to sense the Earth's magnetic field. By using these magnetic sensors, we

Table 7.1 List of sensors used for monitoring the 3D printer and its surrounding environment

Quantity	Modality	Sensor model	Interface	Outputs	Sampling rate (kS/s)	Sensitivity
3	Vibration	ADXL326	Analog	3	1.6 (X,Y), 0.55 (Z)	57 mV/g
3	Magnetic	HMC5883L	Digital (I2C)	3	0.16	2 mG
1	Current	PICO TA018	Analog	1	20	100 mV/Amp
3	Acoustic	AT2021	Analog	1	20	11.2 mV/Pa
1	Acoustic/ vibration	AKG C411 III	Analog	1	18	2 mV/Pa
1	Temperature	LM35	Analog	1	1.5	10 mV/°C
1	Humidity	AM2001	Analog	1	0.0005	0.1% RH

Fig. 7.7 Data acquisition setup for the experimental analysis of digital twin models

measure the fluctuation in the magnetic field of Earth caused by the moving metallic parts of the 3D printer. The sampling rate of the compass is around 270 Hz. The humidity and the temperature of the room change slowly over time. Moreover, the KPIs do not change drastically within the ±5% fluctuation of humidity, and ±1 °C change in the temperature. Hence, any sensor satisfying these properties would be sufficient for monitoring the analog emissions from these modalities.

7.5.1.2 Data Acquisition

NI USB-6343 OEM is used for data acquisition (DAQ), see Fig. 7.7. It has 32 analog inputs. For DAQ, with the increasing number analog inputs the overall sampling rate decreases. This limits the number of analog signals that can be monitored with a high sampling rate. However, for the 25 analog inputs used in our experiment, the resolution of the DAQ is 16 bits for the data with 20,000 Samples/Second sampling rate. These resolutions are sufficient for the acoustic emissions and surpass the requirements for other emissions. Since the DAQ takes an analog signal, an Arduino board coupled with MCP4725 is used to convert the sensor's digital data to analog form. This is done to synchronize all the 25 channels and maintain coherent sampling and data resolution. The Arduino board reads the data from the sensor using the I2C interface and sends it to MCP4725 over the I2C for conversion to analog form. This conversion is necessary for the magnetic sensors. There are three magnetic sensors, each measuring magnetic fluctuation in X, Y, and Z direction. This results in a total of nine signals given by the three magnetic sensors. Since the MCP4725 boards share the same I2C addresses, two TCA9548A I2C Multiplexers are used to access them separately. According to our measurements, this setup can convert more than 6×275 digital samples to analog signal every second, which is more than enough for converting the 3×170 samples generated by the compass sensors.

7.5.1.3 Data Synchronization

The DAQ used in the experiment assures synchronization of all the sensors' data with each other (see Fig. 7.8 for a snapshot of data collected from the DAQ). However, the analog signals collected from the sensors should be mapped with G-code for building the digital twin. For being able to segment the G-code, the 3D printer firmware is modified to send every G-code (along with the time-stamp with accuracy in the range of milliseconds) right before execution to the host's IP address. The port used for communication is 5000. A Python code running on the host side

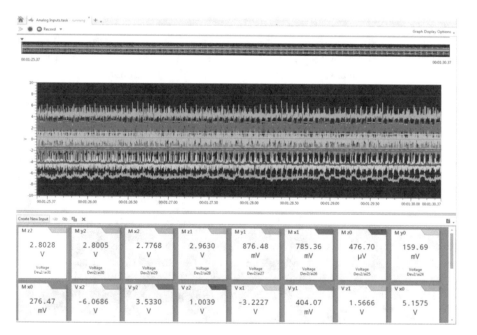

Fig. 7.8 A sample snapshot of the data collected from the sensors

(the desktop computer) is made to continuously listen at port 5000, waiting for the printer to start printing. Once the host receives the first G-code, it first saves the 3D printer's clock data (which allows us to synchronize the G-code with the DAQ data). It then starts saving the sensors data (in *.tdms* format) sent from the DAQ in chunks of ∼55 MB.

7.5.2 Test 3D Objects

For modeling and testing the aliveness of the digital twin, we have customized a benchmark model (see Fig. 7.9). Using this benchmark model, we measure the surface texture (K_1) and dimension variation (K_2), map these KPIs with the corresponding analog emissions from the side-channels, and build the digital twin model. Our benchmark model has four surfaces: top and bottom for the floor and front and back for the wall. Hence, we measure the surface texture of all four surfaces. For dimension, we measure the width and the breadth of the floor, the width and the height of the wall, and the thickness of both the floor and the wall (see Fig. 7.9).

7.5.3 Data Collection

The main objective of the digital twin is to be able to predict the KPIs based on the environmental effects and aging. Collecting analog emissions, within a short period of time, that encompasses the interaction between the system and the environment is a challenging task. Moreover, collecting the variation in analog emissions due to

Fig. 7.9 Test 3D object for digital twin experiment

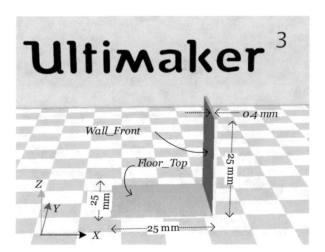

Table 7.2 Summary of environmental and aging degradation parameters

Degradation parameter	Corresponding α and β	Environmental + aging effects	Effects
B_1	Nozzle temperature (α)	(1) Sensor Malfunction, (2) Extreme environmental temperature, (3) Heating element malfunction degradation	K_1, K_2
B_2	Filament feed-rate (α)	(1) Slippage, (2) Worn out rollers due to mechanical degradation, (3) Variation of filament thickness	K_1, K_2
B_3	Acceleration of stepper motors (α)	(1) Rust, (2) Vibration due to loose components, (3) Vibration due to mechanical degradation	K_1, K_2
B_4	Power outage (α)	(1) Short circuit of electronic components, (2) PCB failure, (3) Power supply failure	K_1, K_2
B_5	Printer alignment (α)	(1) Shockwave, (2) Earthquake	K_1, K_2
B_6	Humidity (α)	(1) Faulty HVAC	K_1, K_2
B_7	Slicing thickness (β)	(1) Faulty Z-motor, (2) Erosion of translation screw	K_1, K_2

aging requires collecting data for a long period of time. In order to carry out the experiment within a short period of time, we have performed the following tasks:

- First of all, a summary of environmental and aging effects and their corresponding impacts on the design and process parameters is listed (presented in Table 7.2).
- Analysis of how each design and process parameter gets affected is performed.
- Then design and process parameters are varied to reflect the impact of environmental and aging effects
- The KPIs corresponding to the design and process parameter variation are measured.

From Table 7.2, we were able to analyze which design and process parameters get affected by which environmental factors and aging degradation. For the purpose of the experiment, where we want to validate the DDDAS enabled digital twin, we selected the degradation parameter (B_2) which can be reflected in terms of varying filament flow rate. The flow rate of the system is changed to reflect the effects of the environment and aging. To do this, the flow rate is changed from 20 to 200% of the optimal flow rate of the 3D printer with the step size of 10%. In the 3D printer, the optimal range for the flow rate lies in the range of 80 to 120%, The flow rate is the process parameter which is calculated using the equation:

$$W * H = A = \frac{Q}{v_{feed}} \tag{7.2}$$

where W is the width and H is the height of the line segment being printed on the XY-plane, Q is the constant volumetric flow rate of the material. Q is estimated based on die swelling ration, pressure drop value and buckling pressure of the filament. v_{feed} is the feed velocity of the filament and is calculated as:

$$v_{feed} = \omega_r * R_r \qquad (7.3)$$

where ω_r is the angular velocity of the pinch rollers, and R_r is the radius of the pinch rollers. Based on these values, the pressure drop is calculated as follows:

$$P_{motor} = \frac{1}{2} \triangle P * Q \qquad (7.4)$$

where P_{motor} is the pressure applied by the stepper motors, $\triangle P$ is the pressure drop. Hence, the pressure applied by the motor needs to be maintained for the constant volumetric flow rate. However, this pressure needs to be less than buckling pressure calculated as follows:

$$P_{cr} = \frac{\pi^2 * E * d_f^2}{16 * L_f^2} \qquad (7.5)$$

where E is the elastic modulus of the filament, d_f is the diameter of the filament, and L_f is the length of the filament from the roller to the entrance of the liquefier present in the nozzle. It is evident from these equations that maintaining a constant flow rate depends on various parameters, and environmental or aging factors affecting any of these parameters will change the flow rate, causing changes in the KPIs.

While creating a digital twin in a manufacturing plant, this analysis need not be performed, and data from various analog emissions that have the likelihood of behaving as side-channels can be collected. This data can then be mapped to the KPIs to be able to model a digital twin that predicts the KPIs based on the varying process parameters. The variation in the process parameters is due to the environmental effects and degradation due to aging.

7.5.4 Data Segmentation

The foremost task in any data-driven modeling is acquiring the labeled data for training the machine learning algorithms. In our case, we want to be able to map the KPIs with the corresponding α and β, and the analog emissions. In order to achieve this, we segmented the floor and wall of the test objects into various segments, mapped it to the corresponding G-codes of the object, and acquired the timing data necessary to segment the analog emissions.

For experimental purposes, both the wall and floor region are segmented into 4 by 4 matrix segments. After segmenting the data, the various segments are grouped

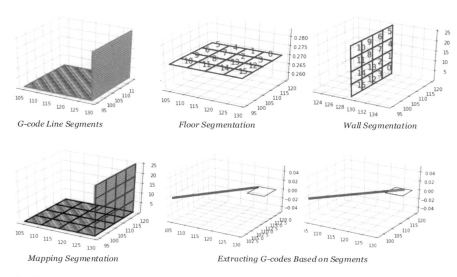

Fig. 7.10 Segmentation of test 3D object for digital twin experiment

Fig. 7.11 Printed test objects

together for the purpose of mapping the corresponding KPIs with analog emissions. For instance, the floor and wall segments numbers 3, 7, 8, and 13 (see Fig. 7.10) are grouped to measure the surface texture and thickness, as they lie on the center region and have homogeneous surface texture and dimension (see Fig. 7.11). The reason for searching homogeneous segments is made clear through Fig. 7.12. Due to the custom synchronizing code running inside the 3D-printer firmware, various non-homogeneous changes are created in the printed 3D object. These changes are more prominent on the outline of the object. Hence for the experimental verification purpose, in the chapter, we have discarded the non-homogeneous outer segments.

For each of the test objects, the total number of channels from which the data collected is 25. And for each channel, the total number of features extracted is $114 + 82 + 58 = 254$. In our training model, all the channel's features are fused together into a single matrix. Hence, the dimension of training data is $X_{samples \times 25*254} = X_{samples \times 6350}$, where samples represent the total number of observations.

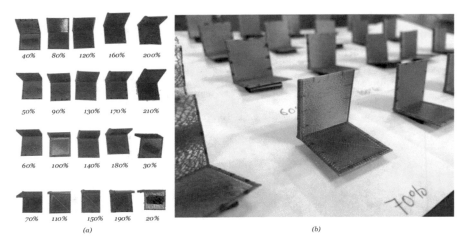

Fig. 7.12 Various test objects printed with varying flow rates. (**a**) Top view. (**b**) Perspective view

7.6   Simulation and Results for Digital Twin Models

7.6.1   Digital Twin Models

With the objective of predicting the dimension and surface texture quality, we have created the following digital twin models.

- Two models, predicting the thickness of the wall and the floor.
- Four models predicting the surface texture of floor top, floor bottom, wall front, and wall back, respectively.

All of these models are created using gradient boosting based regressors [30, 34]. These are an ensemble of decision tree based regression models. By generating a new tree against the negative gradient of the loss function, these algorithms combine weak learners to control over-fitting. It is chosen due to its robustness against outliers and better predictive power against other regression algorithms. In this section, we will present some accuracy measures of the regression algorithm while training it for optimal values of flow rate (80 to 120%), and slowly incorporating degradation along the positive direction, i.e., (130, 140, . . . , 200%).

7.6.2   Aliveness

In order to check the aliveness of the digital twin, we devised the experiment as follows:

1. Train the digital twin with the optimal values of the flow rates (80 to 120%).
2. Assume an environmental degradation has caused the flow rate to vary to 200%. Predict the KPIs, with the digital twin trained with optimal flow rates.
3. Gradually incorporate the flow rates degradation from 130, 140%, etc., all the way to 190%. This incorporation demonstrates a gradual update of the digital twin along with the run-time performance of the 3D printer.

Through this experiment we will be able to measure two things: (1) the prediction capability of the digital twin modeled from side-channel emissions, and (2) cognitive capability enabled for DDDAS by checking the aliveness of the digital twin. The prediction capability is demonstrated in terms of the accuracy of each of the models. By gradually adding the analog emissions in modeling the digital twin, we assume that we are adding the emissions that represent the physical status of the system that is closer to the current state. By doing so, we expect to see improvement in the prediction capability of the digital twin and lower δ value. The lower value of δ will signify that it is possible to monitor the δ and infer about the aliveness of the digital twin model.

Since we have used the gradient boosted trees, the feature re-ranking is performed based on the relative importance of each of the features. It is calculated as follows [30]:

$$\hat{I}_j^2(T) = \sum_{t=1}^{J-1} \hat{i}_j^2 1(v_t = j) \tag{7.6}$$

where we first define a J-terminal node tree T, and sum the result over the non-terminal nodes t. v_t is defined as the splitting variable, and it is associated with each of the node t. The indicator function $1(.)$ has value 1 if its argument is true, and zero otherwise. And $\hat{i}_j^2$ is defined as the estimated empirical improvement in squared error in prediction as a result of split using the particular feature, and it is calculated as:

$$i^2(R_1, R_2) = \frac{w_l w_r}{w_l + w_r}(\bar{y}_l + \bar{y}_r)^2 \tag{7.7}$$

where $\bar{y}_l$, $\bar{y}_r$ correspond to the left and right daughter response means for the node, respectively, and w_l, w_r are the corresponding sums of the weights. For collection of decision trees $\{T_m\}_1^M$ obtained through boosting, Eq. 7.6 can be generalized with an average over all the trees as follows:

$$\hat{I}_j^2 = \frac{1}{M} \sum_{m=1}^{M} \hat{I}_j^2(T_m) \tag{7.8}$$

Hence, using Eq. 7.8, the feature importance is calculated for the boosted trees, and this is used as a metric for re-ranking the features for virtual sensor placement using dynamic data-driven application systems.

The results for modeling the digital twin for predicting the KPIs thickness of the floor and the wall are shown in Figs. 7.13 and 7.14, respectively. The mean absolute error actually represents the mean absolute error for δ. First of all, the accuracy of the digital twin for the optimal flow rate value is around 0.12 mm, which is around the resolution of the 3D printer. This proves that the digital twin can be modeled using the analog emissions which behave as the digital twin. It can also be seen that, as more recent degradation emissions are incorporated in the training sample, the

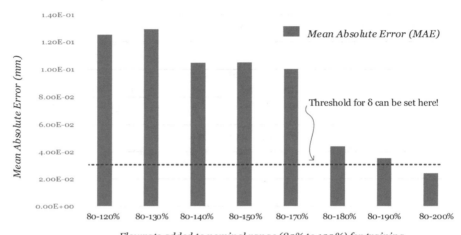

Fig. 7.13 Aliveness test result for digital twin model predicting thickness of the floor

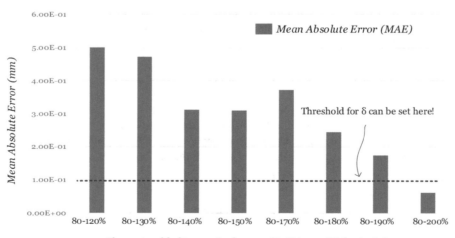

Fig. 7.14 Aliveness test result for digital twin model predicting thickness of the wall

delta value gets smaller and smaller for the digital twin model predicting thickness values for both the surface and the wall.

The delta threshold that can be selected for checking whether the digital twin can be set along ±0.03 mm of the previous delta value for the digital twin predicting the floor thickness. Whereas, for the digital twin predicting the wall thickness, the delta value threshold can be set at ±0.1 mm. This shows the DDDAS system is capable of improving the digital twin model by providing feedback about the aliveness of the model.

The feature re-ranking results for checking whether the digital twin is up-to-date or not for large and small delta value is shown in Figs. 7.15 and 7.16, respectively. From the figure, it can be seen that the feature importance for the digital twin has drastically changed when the delta was large compared to the one with the small delta. This means the DDDAS system is re-ranking the features dynamically and updating the digital twin models. Notice that the feature names are presented as [< Emission Name > _ < Axis of Measurement > _ < Signal Domain) > _ < Feature Name >] for sensors measuring in the three axes, and [< Emission Name > _ < Signal Domain) > _ < Feature Name >] for sensors measuring one-dimensional data (such as temperature, humidity, and power).

The result for checking the results for modeling the digital twin that predict the KPIs surface texture of the floor and the wall is shown in Fig. 7.17a–d, respectively. As predicted the delta value for the digital twin model predicting the surface texture for the bottom surface of the floor, the front surface of the wall, and back surface of the wall are decreasing with the incorporation of more recent analog emissions. The delta threshold that can be selected for checking whether the digital twin can be set along ±0.17 of the previous delta value for the digital twin predicting the top surface's texture of the floor.

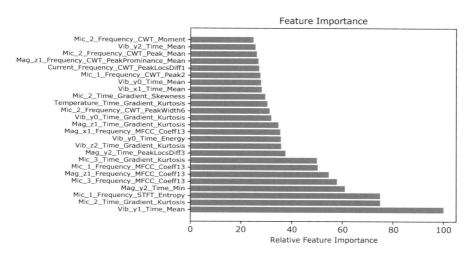

Fig. 7.15 Feature importance for predicting floor thickness with largest delta

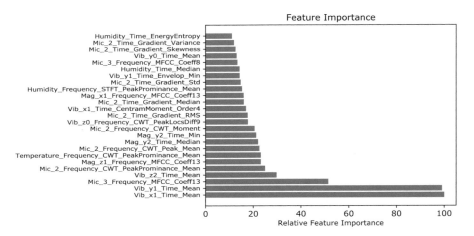

Fig. 7.16 Feature importance for predicting floor thickness with smallest delta

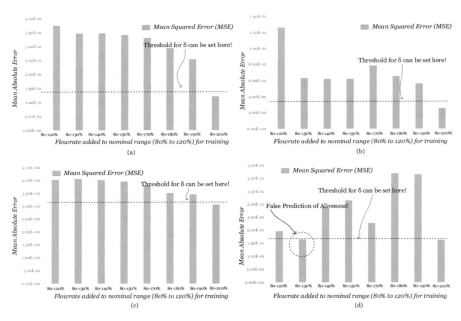

Fig. 7.17 Aliveness test while predicting surface quality. (**a**) Digital twin model for predicting front surface of the wall. (**b**) Digital twin model for predicting front surface of the wall. (**c**) Digital twin model for predicting bottom surface of the floor. (**d**) Digital twin model for predicting top surface of the floor

With this threshold, however, we see that there is one false positive that the digital twin is alive when the flow rate 130% is incorporated. The delta threshold can be set along ±3.5 of the previous delta value for predicting the bottom surface's texture

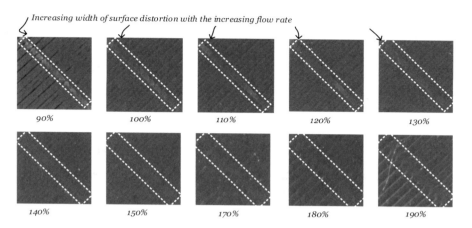

Fig. 7.18 Surface textures with flow rate variation and the case of unaccounted trench

of the floor. The delta threshold can be set along ±0.03 of the previous delta value for predicting the front surface's texture of the wall. The delta threshold can be set along ±0.2 of the previous delta value for predicting the back surface's texture of the wall.

As seen in Fig. 7.17d, checking the aliveness of the digital twin was met with unexpected variation in the delta values. On narrowing down the surface textures, we were able to find the reason for this unexpected behavior. The analysis is presented in Fig. 7.18. In the figure, segment 2's surface of the top surface of the floor is presented for various flow rates. As we can observe that there is a trench-like structure on the surface which has a varying amount of the width for various flow rates (in fact it increases with the flow rate). This is due to the heated nozzle passing very close to the surface of the object. This effect, however, happens on the next layer after the surface has been completed. In our algorithm, we have not incorporated the analog emissions for the layers (G-codes) that are not contributing towards making the surface. Hence, the training algorithms are not able to incorporate these changes, causing it to have unexpected performance.

7.7 Summary

In this chapter, we have presented how dynamic data-driven application systems concepts can be used to re-rank and re-train the digital twin model. This model update is based on measurement of varying δ, which measure the difference in predicted and real key performance indicators (for example, surface texture and dimension). We showed how monitoring δ can enable the digital twin to have a cognitive capability of being aware of its aliveness. This methodology is scalable to create digital twin for multiple key performance indicator predictions, and towards other manufacturing systems as well.

References

1. Wan, J., Canedo, A., & Al Faruque, M. A. (2015). Cyber–physical codesign at the functional level for multidomain automotive systems. *IEEE Systems Journal, 11*(4), 2949–2959.
2. Lee, J., Bagheri, B., & Jin, C. (2016). Introduction to cyber manufacturing. *Manufacturing Letters, 8*, 11–15.
3. Glaessgen, E., & Stargel, D. (2012). The digital twin paradigm for future NASA and US Air Force vehicles. In *Structures, Structural Dynamics and Materials Conference*.
4. Lee, J., Bagheri, B., & Kao, H. A. (2015). A cyber-physical systems architecture for industry 4.0-based manufacturing systems. *Manufacturing Letters, 3*, 18–23.
5. Peng, Y., & Dong, M. (2011). A prognosis method using age-dependent hidden semi-Markov model for equipment health prediction. *Mechanical Systems and Signal Processing, 25*(1), 237–252.
6. Kothamasu, R., Huang, S. H., & VerDuin, W. H. (2006). System health monitoring and prognostics—a review of current paradigms and practices. *The International Journal of Advanced Manufacturing Technology, 28*(9–10), 1012–1024.
7. Parasuraman, R., Sheridan, T. B., & Wickens, C. D. (2000). A model for types and levels of human interaction with automation. *IEEE Transactions on Systems, Man, and Cybernetics-Part A: Systems and Humans, 30*(3), 286–297.
8. Chhetri, S. R., Rashid, N., Faezi, S., & Al Faruque, M. A. (2017). Security trends and advances in manufacturing systems in the era of industry 4.0. In *2017 IEEE/ACM International Conference on Computer-Aided Design (ICCAD)* (pp. 1039–1046). Piscataway: IEEE.
9. Smart Factory Task Group. (2017). Smart factory applications in discrete manufacturing. https://goo.gl/tT9Sf4.
10. Rosen, R., Von Wichert, G., Lo, G., & Bettenhausen, K. D. (2015). About the importance of autonomy and digital twins for the future of manufacturing. *IFAC-PapersOnLine, 48*(3), 567–572.
11. Lee, J., Lapira, E., Bagheri, B., & Kao, H. A. (2013). Recent advances and trends in predictive manufacturing systems in big data environment. *Manufacturing Letters, 1*(1), 38–41.
12. Siemens. (2015). On track for the future – the Siemens Digital Twin show. https://www.youtube.com/watch?v=GzPWBHT1Hl4.
13. IBM. (2017). Introduction to digital twin: Simple, but detailed. https://www.youtube.com/watch?v=RaOejcczPas.
14. Grieves, M. (2014). Digital twin: Manufacturing excellence through virtual factory replication. *White Paper*, 1–7.
15. Bacidore, M. (2015). Digital twin to enable asset optimization. https://www.smartindustry.com/articles/2015/digital-twin-to-enable-asset-optimization/.
16. Kucera, R., Aanenson, M., & Benson, M. (2017). The augmented digital twin. https://goo.gl/RHSwZQ.
17. Parris, C., Laflen, B., Grabb, M. L., Kalitan, D. M. (2017). The future for industrial services: The digital twin. https://www.infosys.com/insights/services-being-digital/Documents/future-industrial-digital.pdf.
18. Marr, B. (2017). What is digital twin technology–and why is it so important. https://www.forbes.com/sites/bernardmarr/2017/03/06/what-is-digital-twin-technology-and-why-is-it-so-important/#77c376b72e2a.

19. NSF I/UCRC Center for Intelligent Maintenance Systems (IMS). (2014). Digital twin for machine monitoring – cyber-physical interface for manufacturing, IMS center. https://www.youtube.com/watch?v=6sh4U44AndQ.

20. Tuegel, E. J., Ingraffea, A. R., Eason, T. G., & Spottswood, S. M. (2011). Reengineering aircraft structural life prediction using a digital twin. *International Journal of Aerospace Engineering, 2011*.

21. IBM. (2017). Creating a building's 'digital twin'. https://www.ibm.com/internet-of-things/iot-zones/iot-buildings/sensors-in-intelligent-buildings/.

22. ANSYS. (2017). Excellence in engineering solutions, advantage: Spotlight on the digital twin. http://www.ansys.com/-/media/Ansys/corporate/resourcelibrary/article/ansys-advantage-digital-twin-aa-v11-i1.pdf.

23. Bacidore, M. (2017). The connected plant enables the digital twin. https://www.controlglobal.com/industrynews/2017/hug-7/.

24. TWI. (2017). TWI embarks on lifecycle engineering asset management through digital twin technology. https://goo.gl/WxLPcy.

25. DebRoy, T., Zhang, W., Turner, J., & Babu, S. S. (2017). Building digital twins of 3D printing machines. *Scripta Materialia, 135*, 119–124.

26. Knapp, G. L., Mukherjee, T., Zuback, J. S., Wei, H. L., Palmer, T. A., De, A., et al. (2017). Building blocks for a digital twin of additive manufacturing. *Acta Materialia, 135*, 390–399.

27. Darema, F. (2004). Dynamic data driven applications systems: A new paradigm for application simulations and measurements. In *International Conference on Computational Science* (pp. 662–669). Berlin: Springer.

28. Hunter, R. S. (1958). Photoelectric color difference meter. *Josa, 48*(12), 985–995.

29. Liu, Z.-Q. (1991). Scale space approach to directional analysis of images. *Applied Optics, 30*(11), 1369–1373.

30. Friedman, J. H. (2001). Greedy function approximation: a gradient boosting machine. *Annals of Statistics*, 1189–1232.

31. Dobra, A., & Gehrke, J. (2002). Secret: A scalable linear regression tree algorithm. In *Proceedings of the Eighth ACM SIGKDD International Conference on Knowledge Discovery and Data Mining* (pp. 481–487). New York: ACM.

32. Maltamo, M., & Kangas, A. (1998). Methods based on k-nearest neighbor regression in the prediction of basal area diameter distribution. *Canadian Journal of Forest Research, 28*(8), 1107–1115.

33. Collins, M., Schapire, R. E., & Singer, Y. (2002). Logistic regression, AdaBoost and Bregman distances. *Machine Learning, 48*(1), 253–285.

34. Pedregosa, F., Varoquaux, G., Gramfort, A., Michel, V., Thirion, B., Grisel, O., et al. (2011). Scikit-learn: Machine learning in python. *Journal of Machine Learning Research, 12*(Oct), 2825–2830.

Chapter 8
IoT-Enabled Living Digital Twin Modeling

8.1 Introduction

In Chap. 7, we presented a digital twin modeling methodology inspired by dynamic data-driven application systems. In this chapter, we will delve deeper into utilizing the side-channels to update the model. Digital twin consists of large historical context and performance data and utilizes the direct (through inbuilt sensors) and indirect (through latent variable analysis) sensing to provide the near real-time representation of the physical system. Moreover, it consists of various models (for simulation, monitoring, control, optimization, etc.,) in a hierarchical manner (consisting of a representation of the system, process, component, etc.,) which can provide the blueprint of the whole system [1].

Cyber-physical additive manufacturing technologies are still susceptible to defects due to the large diversity in the structure and properties of printed components [2]. Digital twin models, in this situation, could alleviate the cost of manufacturing by providing tools to simulate and infer quality deviation in the virtual domain. In this chapter, we narrow down the scope towards fused deposition modeling (FDM) technology based 3D printers, which print 3D objects using thermoplastic filaments such as acrylonitrile butadiene styrene (ABS) or polylactic acid (PLA).

A key enabler for creating a *digital twin* is the availability of a large number of built-in sensors and their historical data. However, current FDM based additive manufacturing printers lack a large number of these sensors [3]. It mainly consists of sensor necessary for basic control (such as a temperature sensor, micro-switch, etc.). Lack of sensor arrays makes it a difficult task for sensing the current system states, which is vital for digital twin models. Moreover, the task of building digital twins becomes even harder once the system has been manufactured as placement and selection of sensors for direct observation of system states can be challenging [4].

© Springer Nature Switzerland AG 2020
S. R. Chhetri, M. A. Al Faruque, *Data-Driven Modeling of Cyber-Physical Systems Using Side-Channel Analysis*,
https://doi.org/10.1007/978-3-030-37962-9_8

Previously, due to the lack of cheap sensors and high-speed and reliable network, acquiring the data for the *digital twin* was costly. Today, thanks to the availability and affordability of IoT sensors, it is becoming easier and cheaper for system operators to acquire large amounts of data on-the-go from their physical systems [5, 6]. However, for 3D printers that do not have built-in sensors and lack historical data, building the *digital twin* with IoT sensors is still a challenge.

8.1.1   Research Challenges

The work presented in this chapter is motivated by three important research questions that apply to legacy FDM technology based additive manufacturing system without built-in sensors and access to historical data:

- Is it possible to build a *digital twin* of a FDM based additive manufacturing system by indirectly monitoring the side-channels?
- Can this indirect side-channel based digital twin model faithfully captures the interaction between the environmental factors, process parameters of the system, and the design parameters of the product to explain the impact of such interaction on products?
- Can we retrofit low-end IoT sensors to maintain the digital twin up-to-date and use it to predict the product quality (localize faults and infer tolerance deviation) of the next product being produced?

8.1.2   Contribution

To address these research challenges, this chapter provides a study on the limits of various sensors modalities (such as acoustic, magnetic, power, vibration, etc.) and their contributions towards building and maintaining a *living digital twin* [7]. The key insight presented in this chapter is that the manufacturing machines generate *unintended* side-channel emissions that carry valuable information about *the machine itself, the product they are producing, and the environment.* Our methodology uses IoT sensors to capture these side-channel information and build a *living digital twin* (see Sect. 8.3.6). Our main contribution is a novel methodology to monitor production machine degradation, build their *living digital twin*, and use this *living digital twin* to provide product quality inference (see Sect. 8.3.7) while localizing faults (see Sect. 8.3.5).

8.1.3 Motivational Case Study for Multi-Sensor Data Analysis

In this chapter, we analyze the data collected from multiple sensors commonly available in IoT devices. These sensors (such as an accelerometer, Hall-effect magnetic sensors, microphone, etc.) are commonly used in the state-of-the-art IoT devices. Moreover, recently more and more sensors (such as current, humidity, temperature, etc.) are added in IoT devices. For building the system *digital twin*, we propose to capture the interaction between the cyber-domain data (such as G/M-codes carrying geometry and process information), the physical domain input of the system (such as raw materials and energy), and the environment. G/M-codes consist of G and M code. The *digital twin* of the product (the 3D object being created) is initially described using a computer-aided design (CAD) tools. The CAD tool then produces stereolithography (STL) files which consist of geometry description of the object in coordinate space. Then a computer-aided manufacturing (CAM) tool takes the STL file and slices it into multiple layers and finds a trace to be followed to print the object in each layer. The output of the CAM layer is the G/M-code.

In our experiment, we consider that the *digital twin* of the product is described using the G/M-code. G-codes are responsible for controlling the motion (XYZ axes and extrusion rate of filaments) of the machine for constructing certain geometry shape. Whereas, M-codes are responsible for controlling the machine parameters (process parameters such as temperature, acceleration of stepper motors, etc.). Recently, researchers in [8, 9] have demonstrated that various signals (such as acoustic, magnetic, etc.) collected from the 3D printer behave as side-channels and reveal information about the cyber-domain. Motivated by these works, we performed a preliminary study to measure the mutual information between the four types of sensor data (*acoustic, magnetic, vibration*, and *power*) and the G/M-codes in a 3D printer. A variation of angle (0° to 90° with step size of 9°) and speed (700 to 3300 mm/min with step size of 100 mm/min) for printing line-segments is encoded in G/M-codes (with total of 297 G/M-codes) and the principal components various time and frequency domain features extracted from the side-channels were used to calculate the mutual information.

Figure 8.1 shows that various data collected from different sensors. The percentage of mutual information in the y axis represents how much of the total Shannon entropy of the G/M-code ($\log_2(297) \approx 8.21$ *bits*) can be explained by the analog emissions. We have assumed the distribution of G/M-codes to be uniform for calculating the Shannon entropy. The figure shows that different sensors have varying level of mutual information with the cyber-domain G/M-codes. This means that they behave as side-channels and provide information about the cyber-domain. However, the data collected from these sensors not only allow us to infer the cyber-domain data, but it also captures the current system status (mechanical degradation, system vibration, effects of the environment on the system, etc.). Hence, the work presented in this chapter leverages the side-channel data to build a living *digital twin*.

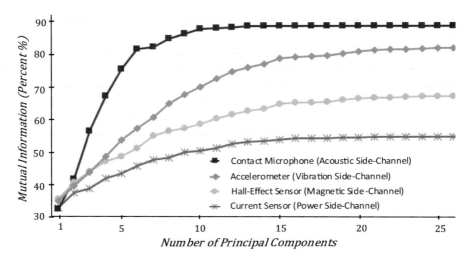

Fig. 8.1 Mutual information analysis on side-channels

8.1.4   Related Work

Since the concept origination, and the onset of emerging technologies, there have
been various efforts to model the *digital twin* of a manufacturing system. Knapp
and Mukherjee in [10] provide building blocks for modeling *digital twin* for laser-
based directed energy deposition additive manufacturing. They use the *digital twin*
to estimate the effects of the process variables on cooling rates, single layer deposit
geometry, and other structural features. Debroy and Zhang in [11] surveyed the
state of the art and motivate the need for more building blocks to create *digital
twins* of additive manufacturing systems. Boschert and Rosen in [12] highlight the
simulation aspect of the *digital twin* and its use in the product life cycle. Alam
and Saddik in [13] provide the reference model for the cloud-based cyber-physical
system with the implementation of a Bayesian belief network for dynamically
updating system based on current contexts. Schroeder and Steinmetz in [14] provide
a methodology to model the attributes related to the *digital twin* for providing easier
data exchange mechanism between the *digital twins*. Cerrone and Hochhalter in
[15] present finite element models of as-manufactured models to predict the crack
path for each specimen. Authors in [16] provide a semantic layer which provides a
mechanism to pass control feedback and evolve the build parameters on-the-fly for
compensating the tolerance.

In summary, all the works have focused on either building the digital twin
using simulation of the first principle-based equations [17, 18] or just placing
expensive sensors for in-situ [19, 20] process monitoring. There is work that uses
low-end sensors for in-situ process monitoring [21–24], however, they do not
consider keeping the model up-to-date, using the indirect side-channels, which

is the fundamental requirement for the *digital twin*. To this end, we propose a methodology for building the system *digital twin* and keeping it alive using low-cost sensors available in off-the-shelf IoT devices. We use the fact that some of the physical emissions act as side-channels, revealing information about cyber-domain, and that for every control signals in cyber-domain, there is a corresponding physical fingerprint in the physical domain. As it uses the side-channels, this methodology is different compared to the existing methods. Using the proposed methodology, we may be able to find new emissions (that may not have been considered during design time) that are able to better represent the cyber and physical states of the system during run-time.

8.2   Background

8.2.1   Concept Definition

As briefly explained earlier, in a manufacturing environment, we have digital and *physical twin* of product and system $DT_{product}$, DT_{system}, $PT_{product}$, and PT_{system}, respectively (see Fig. 8.2). The *digital twin* of the product $DT_{product}$ starts its life cycle in the design phase, where computer-aided design and computer-aided

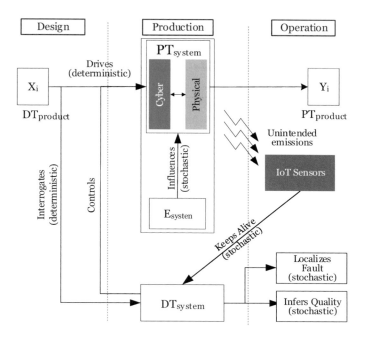

Fig. 8.2 Digital twin concept for manufacturing

manufacturing tools are used to represent the product in the cyber-domain. These product *digital twins* from design phase (with an instance represented using X_i) then go through the production, where the *physical twin* of the system PT_{system} takes raw materials, energy, and the $DT_{product}$ to create its corresponding *physical twins* (with an instance represented using Y_i). The *physical twin* of the system PT_{system} consists of actual physical components that are used for manufacturing. The PT_{system} is influenced by the manufacturing environment in a stochastic manner.

Let $DT_{product} = \{\alpha_1, \alpha_2, \ldots, \alpha_m\} : m \in \mathbb{Z}_{>0}, \alpha \in \mathbb{R}$ represent the parameters that define the *digital twin* of the product (such as dimension, surface roughness, mechanical strength, etc.), $PT_{system} = \{\beta_1, \beta_2, \ldots, \beta_n\} : n \in \mathbb{Z}_{>0}, \beta \in \mathbb{R}$ represent the parameters of the *physical twin* of the manufacturing system (such as flow rate, acceleration values for motors, nozzle temperature, etc.), and let $E_{system} = \{\gamma_1, \gamma_2, \ldots, \gamma_p\} : p \in \mathbb{Z}_{>0}, \gamma \in \mathbb{R}$ represent the environmental factors affecting the manufacturing system (such as temperature, humidity, pressure, etc.). Here m, n, and p represent the total product, system, and environmental parameters that maybe considered for the modeling purpose. We propose to capture the interaction between these parameters using IoT sensors. Using the data collected from multiple modalities (acoustic, vibration, magnetic, power, etc.), we propose to model a stochastic function $\hat{f}(.)$ that performs three tasks: (1) localize the deviation in the $DT_{product}$ parameter from its *physical twin* $PT_{product}$, (2) make sure that the DT_{system} is up-to-date (alive), and (3) infer the quality deviation for the $DT_{product}$ before creating the $PT_{product}$. Moreover, the $DT_{product}$ may interrogate the DT_{system} to infer the quality deviation due to the current status of the PT_{system}.

8.2.2   IoT Sensor Data as Side-Channels

Manufacturing systems consist of cyber and the physical domain. The computing components in the cyber-domain have processes that communicate with the physical domain. A cross-domain signal that is passed from the cyber-domain to the physical domain has the possibility of impacting the physical domain characteristics. This phenomenon is more prominent in manufacturing system where the *digital twin* of the product causes the *physical twin* of the system to behave in a certain deterministic manner. However, due to these characteristics there exists physical emissions (such as acoustic, vibration, magnetic, etc.) which also leak information about the *digital twin* of the product. We denote these emissions as side-channels, as they indirectly reveal the information about the cyber-domain interactions due to the particular physical implementation of the system. For building the *digital twin* of the manufacturing system that captures the interaction between the product *physical twin*, the environment, and the system's *physical twin*, these side-channels play a crucial role in providing the necessary information. As shown in [25, 26], there are various components of the system that reveal information about its internal states through the side-channels. In this chapter, we propose to utilize those indirect side-channel information for fault localization, quality inference, and for updating the digital twin models. In this chapter, we analyze four such analog emissions which potentially behave as side-channels. Let $s_a(t)$, $s_v(t)$, $s_p(t)$, and $s_m(t)$ represent acoustic, vibration, power, and magnetic emissions from the manufacturing system. Then we define each of these signals as:

$$s_a(t) = \hat{\delta}_a(\alpha_i, \beta_j) + \gamma_k : i <= m, j <= n, k <= p \tag{8.1}$$

$$s_v(t) = \hat{\delta}_v(\alpha_i, \beta_j) + \gamma_k : i <= m, j <= n, k <= p \tag{8.2}$$

$$s_p(t) = \hat{\delta}_p(\alpha_i, \beta_j) + \gamma_k : i <= m, j <= n, k <= p \tag{8.3}$$

$$s_m(t) = \hat{\delta}_m(\alpha_i, \beta_j) + \gamma_k : i <= m, j <= n, k <= p \tag{8.4}$$

Equations 8.1–8.4 represent the analog emissions as a result of the deterministic function $\hat{\delta}(.)$ which is influenced by the *digital twin* parameters of the product, $DT_{product}$, and the *physical twin* parameters of the system PT_{system}, and the non-deterministic environmental parameters E_{system}. Moreover, for each of the analog emissions, the total number of parameters (α, β, γ) may not be same. Traditionally, non-trivial simulation based approach such as finite element analysis is used to model the deterministic part and explore relation between the $DT_{product}$, $PT_{product}$, and PT_{system}. However, the PT_{system} parameters vary over time, and E_{system} parameters affect the $PT_{product}$ in a stochastic manner. Hence, we explore the possibility of using IoT sensors to model and maintain a live DT_{system} for product quality inference.

8.2.3 Metric for Quality Measurement

The *digital twin* can be used for various purposes. However, one of the most fundamental uses of *digital twin* is in predicting the key performance indicators (KPIs). Although the ultimate goal of the *digital twin* will be in predicting a variety of KPIs [27], we select quality as one of the KPIs. We will demonstrate that by maintaining a living *digital twin* we can infer the possible deviation in the quality of the product. One of the quality metrics that is used is the dimension (Q_d) of the product.

8.3 Building the Digital Twin

As mentioned earlier, we need to build the *digital twin* from the IoT sensor data to perform three tasks: run-time localization of faults, regularly updating of the system *digital twin*, and inferring the quality of the product *digital twin*.

Hence, in this chapter, the digital twin model consists of algorithms and models associated with fault localization, fingerprint generation, and quality inference. For run-time localization, we propose to create and maintain an active fingerprint library of the individual IoT sensor data corresponding the $DT_{product}$ parameters. This fingerprint also captures the PT_{system} and E_{system} parameters during run-time. Then for localizing the faults, the deviation of the run-time IoT sensor data is compared with the fingerprint. For updating the *digital twin*, a voting scheme is used to check if the majority of the fingerprints are deviating corresponding to a few fingerprints. To infer the deviation in quality, we have proposed to estimate a function $Q_d = \hat{f}(\alpha, \beta, \gamma, s_a(t), s_v(t), s_m(t), s_p(t))$, where the Q_d is a function of $DT_{product}$, PT_{system}, and E_{system} parameters, and the IoT sensor data. The proposed methodology is shown in Fig. 8.3. The various components of the proposed methodology are explained as follows:

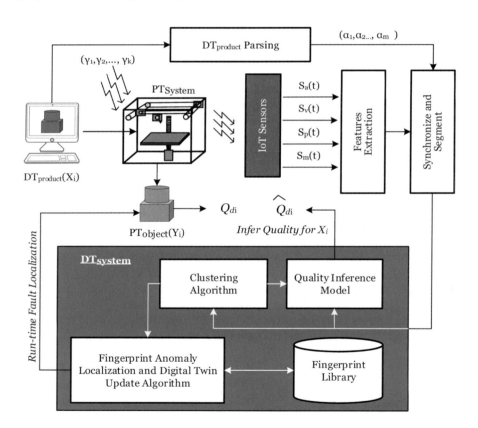

Fig. 8.3 Digital twin modeling methodology

8.3.1 $DT_{product}$ Parsing

For generating the fingerprint of the $DT_{product}$ from the IoT sensor data, first of all it is parsed to its corresponding parameters $(\alpha_1, \alpha_2, \ldots, \alpha_m)$. The parsed values will depend on the type of manufacturing system. In the experimental section, we will present the parsing for an additive manufacturing system that uses G/M-codes. These codes are the instruction that carries the process (machine-specific parameters, such as temperature, acceleration values for motors, etc.) and product information (for example, the geometry description). The parsing will break down the individual parameters from the product *digital twin*.

8.3.2 Feature Extraction

For generating the fingerprint from the analog emissions, various time-domain features such as *energy, energy entropy, peak to peak features* (highest peaks, peak widths, peak prominence, etc.), *root mean square values, skewness, standard deviation, zero crossing rate, Kurtosis* (114 features in total) and frequency domain features such as *mean frequency, median frequency, signal to noise ratio, spectral entropy, spectral flux, spectral roll off* from short term 50 ms time-domain windows (also known as short-term Fourier transform) and continuous wavelet transform (CWT) (140 in total), 20 Mel-frequency cepstral coefficients (MFCCs), etc., are analyzed from IoT sensor data. All the analog signals are first synchronized by performing up and down-sampling and testing the various window size (5–100 ms) for highest model accuracy (50 ms in our case). These features have been selected by calculating the Gini importance or mean decrease impurity of well-known time and frequency domain features (»1000 in total) used for the analysis of time-series data [28]. Principal component analysis (PCA) is then performed to further reduce the dimension of these features. Let m be the total number of reduced feature set, then all the features are concatenated for n total samples to create a feature matrix $O \in \mathbb{R}^{n \times m}$.

8.3.3 Synchronize and Segment

Before clustering is performed, the features are synchronized and segmented into subgroups based on the parsed $DT_{product}$ parameters $(\alpha_1, \alpha_2, \ldots, \alpha_m)$. For instance, the features are segmented based on conditions such as presence or absence of particular component's movement (for example, motors responsible for moving the 3D printer nozzle in X, Y, Z axes). By segmenting based on the parsed parameters, the features are reduced into smaller groups. This allows for further reducing the complexity in acquiring the fingerprints. Henceforth, *group* is used to denote the sub-division of the $DT_{product}$ parameters, which are different than the clusters estimated in the subsequent sections.

8.3.4   Clustering Algorithm

For generating the fingerprint of the parsed parameters of $DT_{product}$, a clustering algorithm is used to generate clusters that group similar features into a single cluster. For analyzing the clustering algorithm and the corresponding fitness of cluster number, the silhouette coefficient is calculated for each sample. It measures the similarity of the feature to its assigned cluster compared to other clusters, with a high value representing its close match to the assigned cluster. It is calculated as follows:

$$silhouette\ coefficient(i) = \frac{b(i) - a(i)}{\max\{a(i), b(i)\}} \tag{8.5}$$

where $a(i)$ is the average intra-cluster distance, and $b(i)$ is the mean of the nearest cluster distance (lowest average distance of i with all other points in another cluster where i is not a member). The clustering is carried out for each group of the features for all the analog emissions. The cluster centers, cluster number, and the corresponding average silhouette coefficient of all the analog emission are stored in a library, effectively representing the fingerprint for the given parsed $DT_{product}$ parameter.

Algorithm 1: Algorithm for the fingerprint library generation for digital twin

---

Input: Features: $O \in \mathbb{R}^{nxm}$, Groups: G, Channels: Ch
Output: Fingerprint: $(G, Ch, $Clusters:$C_k$, Silhouette Scores)

1   Initialize K= $1, 2, \ldots, m$
2   Initialize Silhouette Score Threshold $SC_{Threshold}$
3   Split $O \in \mathbb{R}^{nxm}$ into Test and Train set
4   **foreach** $ch \in Ch$ **do**
5 | **foreach** $g \in G$ **do**
6 | | **foreach** $i \in K$ **do**
7 | | | Estimate i clusters using Train set of Features
8 | | | Use $SC_{Threshold}$ to measure accuracy for clustering the Test Set features
9 | | Select cluster number (k) with highest accuracy
10 | | Re-estimate k cluster with all the Features
11 | | Calculate and Store $Silhouette\ Score_{ch_g}$

12 **return** Fingerprint: $(G, Ch, $Clusters:$C_k$, $Silhouette\ Score_{ch_g})$

The pseudo-code for generating the cluster and saving the fingerprint is presented in Algorithm 1. Features with their corresponding group and channel name are passed as input and the fingerprint in the form of clusters and their corresponding silhouette scores are given as output. First, Lines 1 and 2 initialize the cluster numbers and silhouette score threshold for measuring the accuracy of the cluster estimation. Then Line 3 splits the features into test and train set. Normally 80% of the data is used for training and 20% is used for testing while performing k-fold cross-validation [29] to validate the accuracy. For each cluster number, Line 7 estimates the clusters for the training set. Then, Line 8 measures the accuracy of the estimated cluster with a specified silhouette score threshold for the test set of features. Based on the obtained accuracy in Lines 8, 9–11 select the cluster number, re-estimate the cluster, and store the silhouette scores for all the groups and the channels (acoustic, magnetic, power, and vibration signals).

8.3.5   Anomaly Localization Algorithm

The proposed digital twin model of the system is utilized to detect and localize anomaly in the product. To do this, a fingerprint library is created using Algorithm 1 is used for detecting and localizing the anomalous physical signals corresponding to the $DT_{product}$ while printing. The algorithm for detecting and localizing the deviation from the stored fingerprint is given in Algorithm 2.

Algorithm 2 parses the features of the $DT_{product}$ either run-time or after the product's *physical twin* has been created. Then, using the fingerprint library it estimates the new cluster labels for the parsed features in Line 5. Using these labels and the features the new silhouette coefficient for the parsed features is calculated in Line 6 using Eq. 8.5. If the calculated silhouette coefficient is less than the stored silhouette coefficient $\pm$ threshold $SC_{Threshold}$ then the $DT_{product}$ segment corresponding to the feature is marked as deviating from the previous fingerprint and returned as containing a possible anomaly. Moreover, G/M-code adds layers to print the 3D object in sequential order. Hence, if a fault is detected at a certain time, it can be correlated to locate its position in the 3D object.

8.3.6   Digital Twin Update Algorithm

For updating the *digital twin* model, the library of fingerprint for the $DT_{product}$ has to be updated. However, before updating the library, it should be checked if the anomaly in the fingerprint is temporary or it is due to the degradation of the machine over time. In order to update the *digital twin*, Line 10 in algorithm 2 keep track of all the $DT_{product}$ variables that deviated. Then Line 12 checks if more than $feature_{Threshold}$ of the $DT_{product}$ parameters deviated from the previous fingerprint. Then Line 15 checks if more than $group_{Threshold}$ of the groups deviated

---

Algorithm 2: Algorithm for localizing deviation and checking for digital twin update

---

Input: Features: $O \in \mathbb{R}^{nxm}$, $DT_{product}$
Input: Fingerprint(G, Ch, Clusters:C_k, Silhouette Scores SC_{FP})
Output: Segment of $DT_{product} with deviation$
1 Parse $DT_{product}$ into corresponding parameters
2 Segment Feature into corresponding group
3 **foreach** $ch \in Ch$ **do**
4   **foreach** $g \in G$ **do**
5   Get cluster labels CL_i for Features O_i by assigning features to the nearest cluster in C_k
6   Estimate current silhouette coefficient ($SC_{current}$) for estimated cluster labels
7   **foreach** O *and* CL **do**
8   **if** $SC_{current} < SC_{FP} + SC_{Threshold}$ **then**
9   Store $DT_{product}$ segment (Seg)
10             $DeviationFlag_{ch_g} += 1$
11     $\triangle Deviation_{ch_g} = DeviationFlag_{ch_g}/\text{Total } DT_{product}$
12   **if** $\triangle Deviation_{ch_g} > feature_{Threshold}$ **then**
13         $Deviation_g += 1$
14 $\triangle Deviation_g = Deviation_g/\text{Total Group}$
15 **if** $\triangle Deviation_g > group_{Threshold}$ **then**
16     $Deviation_{ch} += 1$
17 $\triangle Deviation_{ch} = Deviation_{ch}/\text{Total Channel}$
18 **if** $\triangle Deviation_{ch} > channel_{Threshold}$ **then**
19   Use algorithm 1 to update the library
20 **return** Seg

---

from the previous fingerprint. Finally, Line 18 checks if more than $channel_{Threshold}$ of all the channels deviated. If these conditions are met, then in Line 19 the library for the *digital twin* is updated. This threshold for checking the deviation from the fingerprint can be varied for different channels and groups based on the amount of information leaked by each of the side-channels.

8.3.7   *Quality Inference Model*

To infer the quality variation, we estimate a function $Q_d = \hat{f}(., \theta)$, where θ represents a function parameter that needs to be learned. Specifically, we treat Q_d as a function of analog emissions, product design parameters, process parameters, and environmental parameters. The quality deviation occurs due to the fact that the environment (α) affects the PT_{system} process parameters (β). Due to this, when the $DT_{product}$ is sent to the manufacturing system, variations are introduced in the $PT_{product}$. However, when the environment affects (α) the process parameters (β)

it changes the physical structure of various components (for example, creation of rust, mechanical eroding, etc.). These changes may cause the side-channel analog emissions from the PT_{system} to vary. The relations between various environmental factors, process parameters, and design parameters may be modeled using first principle (using physics-based equations). However, estimating such functions will require rigorous multi-domain analysis of the complex mechanical system, and may not reflect variation introduced when the system is operating. Instead, we propose to use a data-driven modeling approach to estimate the function $Q_d = \hat{f}(., \theta)$. This function is estimated using a supervised learning algorithm. To do this, for various α, β, and γ values the corresponding emissions need to be collected. However, for experimental purpose, we assume that the environmental variation affects the β values. Hence, we only vary α and β values and collect the corresponding analog emissions from the side-channels. We extract various time and frequency domain features from these analog emissions and together with α and β, construct a feature matrix $O \in \mathbb{R}^{nxm}$, where m represents the total time and frequency domain features concatenated with α and β parameters, and n represents the total samples. Then, we label each of the rows of $O \in \mathbb{R}^{nxm}$ to its corresponding quality values and use supervised learning algorithm to estimate the function $Q_d = \hat{f}(O \in \mathbb{R}^{nxm}, \theta)$. More specifically, gradient boosting based regressor [30] is used to estimate the function $\hat{f}(.)$. It uses an ensemble of decision trees based regression models. This ensemble generates a new tree against the negative gradient of the loss function and combines weak learner to control over-fitting. Hence, they are robust to outliers and outperform many other learning algorithms as demonstrated in [31]. Since regression trees are used as weak learners, we need to estimate various hyper-parameters such as learning rate, number of weak estimators, maximum depth of the weak learners, etc., to improve the capability of the model to generalize. To do this, the collected feature matrix is divided into test and training set. Then, the testing and training accuracy is used to determine the hyper-parameters that best generalize the function. This estimation function is also updated when the *digital twin* update algorithm reaches a consensus that all the fingerprints are outdated.

8.4 Experimental Setup

8.4.1 IoT Sensors

The experimental setup for modeling the digital twin is shown in Fig. 8.4. For analyzing the analog emissions from the side-channels, four acoustic (AT2021 cardioid condenser and a contact microphone with sampling frequency set at 20 kHz, whereas high-end industrial microphones have higher sampling frequency greater than 40 kHz), one vibration (Adafruit triple-axis accelerometer with output date rate ranging only from 1.56 to 800 Hz and measurement range of up to ±8 g, whereas high-end accelerometers have ranges beyond 1 kHz with measurement

Fig. 8.4 Experimental setup
for modeling the digital twin

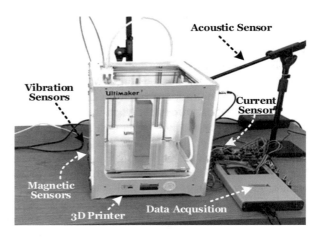

range around ± 50 g), one magnetic (Honeywell's magnetometer HMC5883L with
output date rate ranging from 75 to 160 Hz and measurement range between ± 1 and
± 8 gauss, whereas high-end magnetic field sensors have a date rate range of more
than 1 kHz and measurement range between ± 0.6 and ± 100 gauss, and current
(a low range Pico current clamp with measurement range of 10 mA–20A DC or
rms AC with AC sampling frequency up to 20 kHz with measurement accuracy
of $\pm (6.0\% \pm 30$ mA), whereas high-end sensors have a much smaller resolution of
less than 5 mA in measuring minute current fluctuations) sensors are placed non-
intrusively without hampering the normal operation of the system. In our experiment
for demonstrating the applicability of the proposed methodology, we have used the
above-mentioned sensors which have similar sensor specifications available in IoT
devices [32]. The Cartesian FDM based 3D printer selected for the experiment
is an Ultimaker 3 [33]. The placement of these sensors is performed by position
exploration in Cartesian coordinate. The vibration and magnetic sensors measure
signals in the X, Y, and Z axes. Hence, in total there are four acoustic, three vibration,
three magnetic, and current sensors. We consider them as 11 separate channels.
Analog emissions from the additive manufacturing system (or a 3D printer) are
automatically collected using National Instruments Data Acquisition (NI DAQ)
system whenever a print command is given to it. The analog emission acquisition
was carried out in a lab environment with sound pressure level varying between 60
and 80 dB. The digital twin models are trained and estimated in a desktop computer
with Intel i7-6900K CPU with 3.20 GHz clock frequency, 32 GB of DDR3 RAM,
and 12 GB of NVIDIA Titan X GPU. Moreover, the digital twin models are stored
and retrieved using pickle operation in Python.

8.4.2 Digital Twin Parameters

The sample G/M-code ($DT_{product}$) consists of maximum six parameters, G/M code specifying whether it is machine instruction or coordinate geometry information, travel feed-rate F of the nozzle head, the coordinates in XYZ axes each and amount of extrusion E. Various 3D test objects normally used for calibrating the 3D printer are downloaded from the open-source website [34] to extract sample G/M-codes. The parsing algorithm, in this case, separates the $DT_{product}$ based on presence or absence of 5 of these parameters, G/M, X, Y, Z, and E. Hence, $DT_{product} = \{\alpha_1, \alpha_2, \ldots, \alpha_{32}\}$ and there are 32 groups. When the manufacturing system is operating, the environmental parameters $E_{system} = \{\gamma_1, \gamma_2, \ldots, \gamma_p\}$ affect the *physical twin* parameters $PT_{system} = \{\beta_1, \beta_2, \ldots, \beta_n\}$. This change in β eventually affects the $DT_{product}$ parameters, which in return affects the quality of the product. For example, environmental parameters such as humidity, temperature, etc., may affect the gearbox of the system, which in return may affect the flow rate of the manufacturing system. For experimental purpose changing the environment parameters $E_{system} = \{\gamma_1, \gamma_2, \ldots, \gamma_p\}$ was not performed, instead we have assumed that this parameters eventually affect the β parameter. Hence, analog emissions ($s_a(t)$, $s_v(t)$, $s_m(t)$, and $s_p(t)$) for various α parameters have been collected for optimal β parameters, and the environmental variability has been simulated by varying the β parameters beyond their optimal values to check if the *digital twin* model is able to reflect those changes. After collecting the analog emissions, time and frequency domain features are extracted from each of them. Moreover, the $DT_{product} = \{\alpha_1, \alpha_2, \ldots, \alpha_{32}\}$ consists of timestamps to segment and synchronize the features. For the initial training phase, various G/M-code of the 3D-objects (cube, pyramid, cylinder, etc.) are given to 3D printer and their corresponding analog emissions are collected. From them, we proceed to generate the fingerprint library for maintaining the *digital twin* of the system. Furthermore, in all the training algorithms k-fold cross-validation has been performed to measure the performance of the models and prevent over-fitting.

8.4.3 Sensor Position Analysis

One of the challenges in IoT sensor-based information extraction is figuring out a non-intrusive position of the sensors. This task may also be machine specific. In our experiment, the 3D printers' external surface is considered for non-intrusively placing the sensors. A total of *28* uniform positions are selected as shown in Fig. 8.5. For each of the positions, vibration, acoustic, and magnetic sensors are placed and data is collected for various $DT_{product}$ parameters. Then a gradient boosted random forest is used to create a simple classifier to estimate the accuracy of the model based on various sensor location data. The accuracy of the classifier is given as,

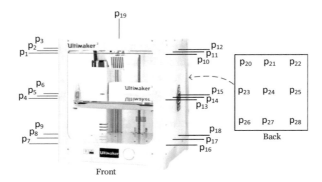

Fig. 8.5 Experimental setup for sensor position exploration

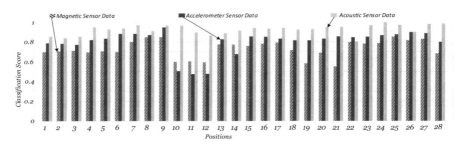

Fig. 8.6 Classification accuracy score for sensors positions

$$Accuracy = \frac{TP + TN}{TP + TN + FP + FN} \tag{8.6}$$

where TP stands for total true positives, TN stands for total true negatives, FP stands for total false positives, and FN stands for total false negatives. The accuracy presented in Eq. 8.6 is taken as a metric for determining the placement of the sensors around the 3D printer. $DT_{product}$ parameters selected for estimating the classifier consists of simple G/M-code instructions (such as presence or absence of stepper motors movement in X, Y, and Z axes).

Accuracy score of IoT sensor data is shown in Fig. 8.6, this score shows which of the positions of the sensors are capable of better classifying the stepper motor movements. It may be noticed that for different positions the classification accuracy is different. Moreover, these accuracy results also correlate the mutual information between the various sensor positions and the side-channels themselves. Based on these values, a single position is selected for each of the sensors. However, since four acoustic sensors are used, positions with top four classification accuracy are selected for the sensor placement.

Fig. 8.7 Scatter plots of the clusters plotted for acoustic side-channel

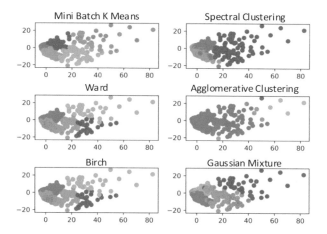

8.4.4 Performance of Clustering Algorithms

Various algorithms are explored for creating clusters to generate the fingerprints. Among them are *mini batch K-means*, *spectral clustering*, *ward*, *agglomerative clustering*, *birch*, and *Gaussian mixture method*. For each of the clustering algorithm, a varying number of clusters are initialized, and the corresponding silhouette coefficient is calculated for measuring the fitness of the features into these clusters.

The average silhouette coefficients of the clustering algorithms for all groups and channels are shown in Fig. 8.8, and the corresponding scatter plots of acoustic side-channel for cluster number five are shown in Fig. 8.7. It may be noticed that although the *agglomerative clustering* has a higher silhouette coefficient value, from the scatter plot, the clusters are not well distributed in the scatter plot. However, the *birch* clustering algorithm has relatively higher silhouette coefficient with a better spread of the cluster centers. Hence, *Birch* algorithm is selected for generating the clusters for fingerprinting the $DT_{product}$. Furthermore, the number of clusters is also estimated based on the accuracy of the *Birch* algorithm using Algorithm 1.

8.4.5 Anomaly Localization Accuracy

For testing the accuracy of the *digital twin* for detecting the anomalous signals that can possibly cause deviation in the quality of the product, specialized test 3D object is designed (see Fig. 8.9). We have simulated variability of the environment by varying one of the $PT_{system} = \{\beta_1, \beta_2, \ldots, \beta_n\}$ parameters. In our experiment, we have selected flow rate as one of the β parameters. Flow rate should be maintained for uniform deposition of the filament while printing in fused deposition modeling based 3D printers. However, due to sudden slippage, faulty filament, etc., the flow of the filament may deviate from its nominal value.

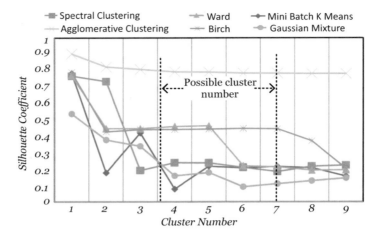

Fig. 8.8 Silhouette coefficient of clustering algorithms

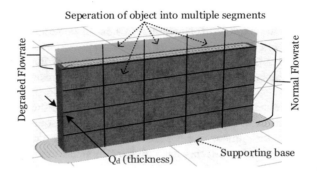

Fig. 8.9 Test object for testing anomaly localization

The flow rate, a process specific parameter, is calculated as follows:

$$W * H = A = \frac{Q}{v_{feed}} \tag{8.7}$$

where W is the width and H is the height of the line-segment being printed on the XY-plane, Q is the constant volumetric flow rate of the material. Q is estimated based on die swelling ration, pressure drop value, and buckling pressure of the filament. And $v_{feed} = \omega_r * R_r$ is the feed velocity of the filament. Where ω_r is the angular velocity of the pinch rollers, and R_r is the radius of the pinch rollers. Then, the pressure drop is calculated as follows:

$$P_{motor} = \frac{1}{2} \triangle P * Q \tag{8.8}$$

where P_{motor} is the pressure applied by the stepper motors, $\triangle P$ is the pressure drop. Hence, the pressure applied by the motor needs to be maintained for the constant volumetric flow rate. This pressure needs to be less than buckling pressure which is calculated as follows:

$$P_{cr} = \frac{\pi^2 * E * d_f^2}{16 * L_f^2} \tag{8.9}$$

where E is the elastic modulus of the filament, d_f is the diameter of the filament, and L_f is the length of the filament from the roller to the entrance of the liquefier present in the nozzle. A sudden change in the pressure can cause the uniform flow of the filament to be disrupted. For validating the application of the digital twin in anomaly localization, the flow rate is varied outside the optimal range (<80 and $>120\%$) at a specific location (see Fig. 8.9) for multiple 3D objects. The anomalous flow rate variation introduced is between 40 and 180% with the step size of $\pm 10\%$. Then the *digital twin* is tested to see if can accurately classify the deviation in quality as an anomaly at the specific location. This is done by first segmenting the 3D object and assigning labels (1 for anomalous flow rate outside the optimal range, and 0 for normal flow rate) to these segments. Then comparing these labels with the results of the Algorithm 2.

The *digital twin* consists of fingerprints for the optimal flow rate in its library, and the corresponding clusters of the individual channels. When the object is printed, the corresponding features are passed to the *digital twin*, and the silhouetted coefficients corresponding the $DT_{product}$ is calculated. Based on Algorithm 2, the analog emissions in each channel is labeled as either being within the deviation limit or exceeding the deviation limit of the silhouette coefficient. For selecting the optimal threshold for making this decision, initially the threshold is varied and the corresponding accuracy of the detection mechanism is measured.

Based on the highest accuracy acquired for each of the channels, the threshold is set and the corresponding classification accuracy for the segments that have been degraded is calculated. Corresponding to the varying threshold the receiver operating characteristic (ROC) curve for some of the channels is presented in Fig. 8.10. The accuracy of each channel in detecting the anomalous flow rate is shown in Fig. 8.11. Since the features are time stamped, the corresponding section of the $DT_{product}$ maybe calculated after the *digital twin* has marked the features to be anomalous. From Fig. 8.11, it can be seen that analog emissions from microphone number four are more accurate in detecting the degradation of the flow rate. This is due to the fact that this emission is collected by the contact microphone attached near the extruder's stepper motor. Moreover, the average accuracy across all the channels in detecting the anomalous flow rate is **83.09%**.

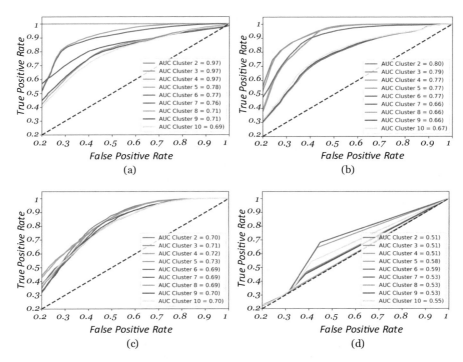

Fig. 8.10 Average receiver operating characteristic curve for anomaly localization. (**a**) ROC curve for sensor data from Mic_4. (**b**) ROC curve for sensor data from Vib_x. (**c**) ROC curve for sensor data from Mag_y. (**d**) ROC curve for sensor data from current

8.4.6   System Degradation Prediction Analysis

For detecting the degradation of the system, and hence the need for updating the *digital twin*, the flow rate for the entire $DT_{product}$ is varied beyond the optimal range. From Eqs. 8.7 to 8.9, it is evident that various mechanical degradation (such as worn out rollers), stepper motor degradation over time, etc., may cause the flow rate to be reduced over time.

To check if the digital twin model gets updated to reflect the current status of the system, we perform two experiments. In the first experiment, the current *digital twin* with its fingerprint library is used to predict the class labels (*True* for update and *False* for do not update) for the degraded flow rate (60%). Then based on the result of Algorithm 2, the updated (or the old) *digital twin* is used to predict the class labels again for the same degraded flow rate (60%) to see if the *digital twin* gets updated again. The result of degradation analysis is presented in Table 8.1.

Table 8.1 consists of *true negative rate, false positive rate* and update decision taken for each channel for the old cluster. When the system degrades, we expect the *digital twin* to find higher negative labels being generated as the silhouette score will

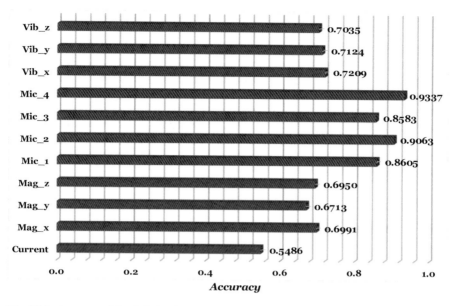

Fig. 8.11 Accuracy of the digital twin's anomaly localization

Table 8.1 Degradation test result for the digital twin

	Old Clusters			New Clusters		
Channel	TNR	FPR	Update	TPR	FNR	Update
Mic_1	0.97423	0.0257	True	0.6040	0.3960	False
Mic_2	0.4962	0.5038	False	0.7460	0.2540	False
Mic_3	0.9705	0.0295	True	0.5731	0.4269	False
Mic_4	0.9867	0.0133	True	0.9798	0.0202	False
Current	0.5924	0.4076	True	0.5545	0.4455	False
Vib_x	0.9324	0.0676	True	0.8681	0.1319	False
Vib_y	0.9695	0.0305	True	0.7224	0.2776	False
Vib_z	0.4602	0.5398	False	0.4400	0.5600	True
Mag_x	0.4791	0.5210	False	0.6718	0.3282	False
Mag_y	0.4382	0.5618	False	0.4344	0.5656	True
Mag_z	0.6267	0.3733	True	0.3669	0.6331	True

be lower than the average silhouette score stored for all the channels and groups. It can be seen that out of eleven channels four of them had the decision of not updating the cluster, and seven of them opting for updating the clusters. Hence, the clusters are updated by Algorithm 2. On the other hand, once the cluster has been updated, the analog emissions are labeled as true, hence we expect to see a higher true positive rate and lower false negative rate. In the Table 8.1, it can be seen that only three of the channels gave the decision for updating the clusters again, however, eight of them opted for not updating the cluster. This shows that the *digital twin* is able to update itself during degradation that causes emissions in multiple side-channels to vary.

It may be noted that the side-channels gave different decisions for updating the digital twin. Out of them, the acoustic sensors and vibration side-channels were mostly able to predict the right decision, whereas the magnetic sensors were mostly wrong in this decision. This also correlates with the accuracy values presented in Fig. 8.11. One anomaly to this is the current sensor data. However, it may be noticed that during both the decision model's true positive and true negative rates are very low compared to the acoustic and magnetic sensors. This means that the current side-channel data is not as reliable for the update decision as to the acoustic and the vibration side-channels.

8.4.7   Quality Inference

For checking the accuracy of the *digital twin* in inferring the deviation of quality (Q_d), first of all, gradient boosting based ensemble of regressors is used to estimate the function $Q_d = \hat{f}(.)$ for the optimal flow rate range (80 to 120%). For each flow rate, five test objects (with the thickness of 4 mm for $DT_{product}$) are 3D printed, and for each test object, various segments (see Fig. 8.12) are created to measure the thickness using the micrometer. Then all the groups of features lying in these segments are assigned a single thickness value. Initially function $Q_d = \hat{f}(.)$ is estimated using optimal flow rate. Then it is used to infer the thickness of the 3D object for various feature samples with varying flow rates. The accuracy of the $DT_{systems}$ quality inference model is measured using mean absolute error value.

The result of the quality inference is shown in Fig. 8.13. At first, the mean absolute error value of the inference model trained with optimal flow rate range is measured. It can be seen in the figure that for optimal flow rate ranges, the mean absolute error value is around 0.5 mm. Then, at each consecutive step the flow rate of the Pt_{system} is varied with a step size of $+10\%$ in the positive direction ($>120\%$)

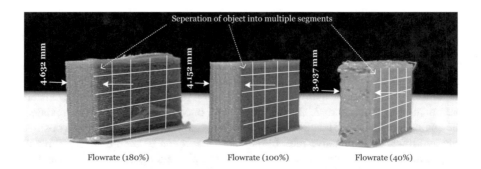

Fig. 8.12 $PT_{product}$ created for testing quality inference and update capability

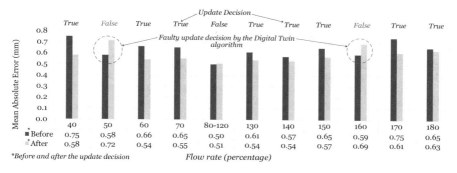

Fig. 8.13 Accuracy of the quality inference model

and at the same time +10% in the negative direction (<80%). It may be seen that in both directions when the system ages (degrades with an increase or decrease in the flow rate), the DT_{system} has increased mean absolute error without the update. This is intuitive as the DT_{system} has not been updated to the new fingerprints. However, once it has been updated the mean absolute error is lower. It may also be noticed that when the system degraded with flow rates at 160 and 50%, the wrong decision was taken by the Algorithm 2 in not updating the quality inference model. Due to this, a large increase in mean absolute error was observed for quality prediction other $DT_{product}$. However, this faulty decision was recovered in the consecutive stages. Moreover, the average mean absolute error in predicting the quality was 0.59 mm (calculated by averaging the mean absolute errors of the inference model after update decision).

8.4.8 Comparative Analysis

Although we present a novel methodology of building a living *digital twin* by using IoT based sensors, there has been a considerable amount of work in quality prediction in additive manufacturing or manufacturing systems in general. In this section, we provide a qualitative comparative study of the various non-exhaustive list of methods compared to the proposed methodology. The result of the comparison is shown in Table 8.2. It may be observed that there are three general categories of research effort in maintaining quality.

The first is the first principle-based approach (simulation) [17, 35], where quality inference model is based on the process and design parameters. These models although are accurate, they do not account system degradation over time and requires a non-trivial formulation of physics-based equations. The second category involves in-situ process monitoring methodologies [36, 37]. These methods monitor the process variation using high-end acoustic and piezoelectric sensors. Compared to these high-end sensor-based methods, our method is able to keep the model

Table 8.2 Comparative analysis of the proposed methodology

Work/system	Method	Metric	Sensors	Anomaly detection accuracy	Checks model update	Quality inference accuracy
[21]/FFF	Bayesian DP mixture model	Build failure detection	Accelerometer, thermocouple, IR, borescope	85% (Average F-score)	×	–
[22]/FDM	Functional qualitative quantitative model	Dimension, surface	Accelerometer, IR, thermocouple	–	×	~0.3 mm (median RMSPE)
[36]/FDM	Support vector machines	Abnormal extrusion detection	High-end acoustic sensor	95%	×	–
[35]/FDM	First principle model of filament	Dimension	–	–	×	~0.1 mm MAE
[23]/FFF	Online sparse estimation-based classification	Abnormal extrusion detection	Accelerometer, IR, thermocouple	90% (F-score)	×	–
[38]/FDM	Hidden semi-Markov model	Abnormal extrusion detection	High-end acoustic sensor	91.9% (Accuracy rate)	×	–
[17]/FDM	Theoretical model	Surface	–	–	×	5.66% MAPE
[20]/SLM	Spectral convolutional neural networks	Build quality	High-end acoustic sensor	79–84%	×	–
[37]/ FDM	Heterodyne technique	Belt fault detection	High-end acoustic and piezoelectric sensor	–	×	–
QUILT/FDM	Behavioral modeling (random forest, clustering)	Dimension	Low-end acoustic, accelerometer, magnetic, current	83.09% (Classification score)	✓	0.59 mm MAE

FDM fused deposition modeling, *SLM* selective lase melting, *MAE* mean absolute error, *FFF* fused filament fabrication, *RMSPE* root mean square percentage error, *MAPE* mean absolute percentage error

updated even using low-end sensor data for fault localization and quality inference.[1] The third category involves process monitoring using low-end sensor placement [21–23]. They focus either on specific anomaly detection or quality variation detection. However, these methods do not consider checking the aliveness (up-to-date model) of the model and are mostly limited to anomaly detection. Each of these techniques has its own merit, hence, the proposed methodology is not intended to function independently but in conjunction with various approaches to fully realize the concept of *digital twin*.

8.5 Discussion

Quality Inference To validate the proposed methodology, the flow rate was used to detect anomalous system behavior and overall system degradation behavior. However, there can be multiple PT_{system} parameters that might affect the quality. However, our methodology can be adjusted over time to consider variation in other PT_{system} parameters over time. We have considered only dimension (thickness of a simple 3D object) as a quality metric. However, for building the full scale DT_{system}, multiple metrics are needed to be considered.

More IoT Sensors and Placement In this chapter, sensors with a low sampling rate and resolution were used. The number of sensors was limited as well. To improve the accuracy of the *digital twin* techniques such as [3] needs to be incorporated for the development of IoT sensor arrays.

Implementation Using IoT Device For building the DT_{system} using IoT devices, further consideration is required for off-the-shelf and wireless IoT devices [3]. Hence, further analysis is required to understand the trade-off between power, time, and performance of the DT_{system} in localizing and inferring the quality variation.

More Test Cases One of the limitations of the experimental section was in using a limited number of test 3D objects for inferring the quality and localizing the faults. However, these 3D objects contain structures which provide large possible variation in G/M-code for building the digital twin models. Nonetheless, digital twin models will be more accurate if large test data are incorporated.

Sensor Fusion Analysis In the motivation section, we provided mutual information analysis for individual side-channels. We acknowledge that a rigorous analysis of the calculation of mutual information by fusing these sensors may further justify the proposed methodology. The machine learning algorithm achieves this by carefully selecting the features extracted from the side-channels to build a model

---

[1]With high-end sensors as theirs, our methodology may achieve higher accuracy in anomaly detection along with the capability of keeping the digital twin most up-to-date at the cost of more computational and resource requirements, which may not be feasible for an IoT paradigm.

Table 8.3 Other additive manufacturing technologies

	Source of analog emissions		
Technology	Acoustic	Vibration	Power
SLA	Build platform, stepper motor	Sweeper	Motor controller
SLS	Fabrication piston	Rollers	Power supply
MJ	Build tray, jetting head	Moving head, blower, position belt	Heater, coil
SLM	Retractable platform	Leveling cylinder	Power supply
EBM	Build platform	Build platform	High voltage cable

SLA stereolithography, *SLS* selective laser sintering, *MJ* material jetting, *SLM* selective laser melting, *EBM* electron beam melting

to get the highest possible accuracy. Future work will require a deeper analysis of mutual information to further clarify the contribution of individual side-channels in indirectly building the digital twin.

Generalizability of the Proposed Method As a case study, we presented applicability of the proposed methodology in fused deposition modeling based additive manufacturing. However, many other technologies also have analog emissions (see Table 8.3), which may be capable of aiding in the indirect method of building digital twin models. Hence, we hypothesize that the proposed methodology will be able to scale across multiple manufacturing systems.

8.6   Summary

In this chapter, we presented a novel methodology to build a living *digital twin* of the fused deposition modeling technology based additive manufacturing system by utilizing various retrofitted low-end sensors available in IoT devices to indirectly monitor the system through various side-channels (such as acoustic, vibration, magnetic, and power). Based on these signals, a clustering algorithm is used to generate a fingerprint library that effectively represents the physical status or the *physical twin* of the system. The *digital twin* is used for localizing the anomalous physical emissions that have the potential of resulting in quality variation. For localizing the error, the *digital twin* achieved an average accuracy of **83.09%**. Moreover, we also presented an algorithm for updating the *digital twin* and inferring the quality deviation. As a case study the *digital twin* modeling was performed on the additive manufacturing system. Compared to the state-of-the-art methods (which do not consider model aliveness),the methodology presented in this chapter is able to update itself, infer quality deviation, and localize anomalous faults in the cyber-physical additive manufacturing system.

References

1. Grieves, M., & Vickers, J. (2017). Digital twin: Mitigating unpredictable, undesirable emergent behavior in complex systems. In *Transdisciplinary perspectives on complex systems*. Berlin: Springer.
2. Mukherjee, T., & DebRoy, T. (2019). A digital twin for rapid qualification of 3d printed metallic components. *Applied Materials Today, 14*, 59–65.
3. Baumann, F., Schön, M., Eichhoff, J., & Roller, D. (2016). Concept development of a sensor array for 3d printer. *Procedia CIRP, 51*, 24–31.
4. He, Y., Guo, J., & Zheng, X. (2018). From surveillance to digital twin: Challenges and recent advances of signal processing for industrial internet of things. *IEEE Signal Processing Magazine, 35*(5), 120–129.
5. Banafa, A. (2016). IoT standardization and implementation challenges. In *IEEE internet of things newsletter*.
6. Petrenko, A. S., Petrenko, S. A., Makoveichuk, K. A., & Chetyrbok, P. V. (2018). The IIoT/IoT device control model based on narrow-band IoT (NB-IoT). In *Young researchers in electrical and electronic engineering (EIConRus)*. Piscataway: IEEE.
7. Chhetri, S. R., Faezi, S., Canedo, A., & Faruque, M. A. A. (2019). Quilt: Quality inference from living digital twins in IoT-enabled manufacturing systems. In *Proceedings of the International Conference on Internet of Things Design and Implementation* (pp. 237–248). New York: ACM.
8. Song, C., Lin, F., Ba, Z., Ren, K., Zhou, C., & Xu, W. (2016). My smartphone knows what you print: Exploring smartphone-based side-channel attacks against 3D printers. In *Proceedings of the 2016 ACM SIGSAC Conference on Computer and Communications Security*. New York: ACM.
9. Hojjati, A., Adhikari, A., Struckmann, K., Chou, E., Tho Nguyen, T. N., Madan, K. (2016). Leave your phone at the door: Side channels that reveal factory floor secrets. In *Proceedings of the 2016 ACM SIGSAC Conference on Computer and Communications Security*. New York: ACM.
10. Knapp, G. L., Mukherjee, T., Zuback, J. S., Wei, H. L., Palmer, T. A., De, A., et al. (2017). Building blocks for a digital twin of additive manufacturing. *Acta Materialia, 135*, 390–399.
11. DebRoy, T., Zhang, W., Turner, J., & Babu, S. S. (2017). Building digital twins of 3D printing machines. *Scripta Materialia, 135*, 119–124.
12. Boschert, S., & Rosen, R. (2016). Digital twin-the simulation aspect. In *Mechatronic Futures*. Berlin: Springer.
13. Alam, K. M., & El Saddik, A. (2017). C2PS: A digital twin architecture reference model for the cloud-based cyber-physical systems. *Access, 5*, 2050–2062.
14. Schroeder, G. N., Steinmetz, C., Pereira, C. E., & Espindola, D. B. (2016). Digital twin data modeling with automationML and a communication methodology for data exchange. *IFAC-PapersOnLine, 49*(30), 12–17.
15. Cerrone, A., Hochhalter, J., Heber, G., & Ingraffea, A. (2014). On the effects of modeling as-manufactured geometry: Toward digital twin. *International Journal of Aerospace Engineering, 2014*.
16. Garanger, K., Feron, E., Garoche, P. L., Rimoli, J. J., Berrigan, J. D., Grover, M., et al. (2017). Foundations of Intelligent Additive Manufacturing. arXiv preprint:1705.00960.
17. Di Angelo, L., Di Stefano, P., & Marzola, A. (2017). Surface quality prediction in FDM additive manufacturing. *The International Journal of Advanced Manufacturing Technology, 93*(9–12), 3655–3662.
18. Boschetto, A., Bottini, L., & Veniali, F. (2016). Integration of FDM surface quality modeling with process design. *Additive Manufacturing, 12*, 334–344.
19. Shevchik, S.A., Kenel, C., Leinenbach, C., & Wasmer, K. (2018). Acoustic emission for in situ quality monitoring in additive manufacturing using spectral convolutional neural networks. *Additive Manufacturing, 21*, 598–604.

20. Wasmer, K., Kenel, C., Leinenbach, C., & Shevchik, S. A. (2017). In situ and real-time monitoring of powder-bed AM by combining acoustic emission and artificial intelligence. In *International Conference on Additive Manufacturing in Products and Applications*. Berlin: Springer.
21. Rao, P. K., Liu, J. P., Roberson, D., Kong, Z. J., & Williams, C. (2015). Online real-time quality monitoring in additive manufacturing processes using heterogeneous sensors. *Journal of Manufacturing Science and Engineering, 137*(6), 061007.
22. Sun, H., Rao, P. K., Kong, Z. J., Deng, X., & Jin, R. (2018). Functional quantitative and qualitative models for quality modeling in a fused deposition modeling process. *Transactions on Automation Science and Engineering, 15*(1), 393–403.
23. Bastani, K., Rao, P. K., & Kong, Z. (2016). An online sparse estimation-based classification approach for real-time monitoring in advanced manufacturing processes from heterogeneous sensor data. *IIE Transactions, 48(7)*, 579–598.
24. Chhetri, S. R., Faezi, S., & Faruque, M. A. (2017). Digital twin of manufacturing systems. *CECS technical report CECS TR-17-07*. http://cecs.uci.edu/files/2018/03/cecs_tech.pdf.
25. Faruque, A., Abdullah, M., Chhetri, S. R., Canedo, A., & Wan, J. (2016). Acoustic side-channel attacks on additive manufacturing systems. In *International Conference on Cyber-Physical Systems*. Piscataway: IEEE.
26. Faezi, S., Chhetri, S. R., Malawade, A. V., Chaput, J. C., Grover, W. H., Brisk, P., et al. (2019). Oligo-snoop: A non-invasive side channel attack against DNA synthesis machines. In *Network and distributed system security symposium (NDSS)*
27. Zhu, L., Johnsson, C., Mejvik, J., Varisco, M., & Schiraldi, M. (2017). Key performance indicators for manufacturing operations management in the process industry. In *Industrial engineering and engineering management (IEEM)*. Piscataway: IEEE.
28. Christ, M., Kempa-Liehr, A. W., & Feindt, M. (2016). Distributed and parallel time series feature extraction for industrial big data applications. arXiv preprint:1610.07717.
29. Refaeilzadeh, P., Tang, L., & Liu, H. (2009). Cross-validation. In *Encyclopedia of database systems*. Berlin: Springer.
30. Elith, J., Leathwick, J. R., & Hastie, T. (2008). A working guide to boosted regression trees. *Journal of Animal Ecology, 77*(4), 802–813.
31. Chhetri, S. R., Canedo, A., & Al Faruque, M. A. (2016). KCAD: Kinetic cyber-attack detection method for cyber-physical additive manufacturing systems. In *Proceedings of the 35th International Conference on Computer-Aided Design*. New York: ACM.
32. Postscapes. (2018). IoT Sensors and Actuators. https://www.postscapes.com/trackers/video/the-internet-of-things-and-sensors-and-actuators/.
33. Ultimaker. (2018). Ultimaker 3. https://ultimaker.com/en/products/ultimaker-3.
34. Thingiverse. (2017). https://www.thingiverse.com/.
35. Boschetto, A., & Bottini, L. (2014). Accuracy prediction in fused deposition modeling. *The International Journal of Advanced Manufacturing Technology, 73*(5–8), 913–928.
36. Wu, H., Wang, Y., & Yu, Z. (2016). In situ monitoring of FDM machine condition via acoustic emission. *The International Journal of Advanced Manufacturing Technology, 84*(5–8), 1483–1495.
37. Yoon, J., He, D., & Van Hecke, B. (2014). A PHM approach to additive manufacturing equipment health monitoring, fault diagnosis, and quality control. In *Prognostics and Health Management Society Conference*. Citeseer.
38. Wu, H., Yu, Z., & Wang, Y. (2017). Real-time FDM machine condition monitoring and diagnosis based on acoustic emission and hidden semi-Markov model. *Journal of Advanced Manufacturing Technology, 90*(5–8), 2027–2036.

Part IV
Non-Euclidean Data-Driven Modeling of Cyber-Physical Systems

Chapter 9
Non-euclidean Data-Driven Modeling Using Graph Convolutional Neural Networks

9.1 Introduction

A system design process consists of various steps such as problem definition, background research, requirement specification, brainstorming solutions, selecting the best solution, developing a prototype, testing, and finally redesigning [1]. During each of these steps, a wide variety of high volume and continuous data is generated. These design steps are repeated throughout various systems such as mechanical, electronic, software, etc. Due to the repetitive nature of these tasks, engineers can save a large amount of time if they can compare existing similar designs that closely match the desired functionality while designing a new system. Rather than having to redesign, they can find functionally similar designs and modify such designs to fit their needs. Moreover, this can further lead to the creation of artificial intelligent assistants that assist human experts to design new systems faster.

The engineering design data varies from domain to domain. In electronic design, it consists of high-level design descriptions, register transfer level descriptions in Verilog or VHDL, schematics, etc. In mechanical design, it consists of data regarding structural designs, modeling, and analysis of components, etc. Moreover, there is a wealth of data generated throughout the supply chain of engineering including computer-aided design (CAD) and computer-aided manufacturing (CAM) tools. In order to perform meaningful learning from this data, we need to utilize non-euclidean or graph learning algorithms that are able to extract, categorize, and label these sparse data.

In this chapter, we propose to utilize a structural graph learning algorithm to abstract the detailed engineering data (such as configuration, code, hybrid equations, geometry, sensor data, etc.). This chapter expands the general view presented in [2]. To achieve this, we first represent the engineering data using a knowledge graph and then perform semi-supervised learning to be able to classify the sub-graphs based on their structural property and the corresponding attributes. This will

© Springer Nature Switzerland AG 2020
S. R. Chhetri, M. A. Al Faruque, *Data-Driven Modeling of Cyber-Physical Systems Using Side-Channel Analysis*,
https://doi.org/10.1007/978-3-030-37962-9_9

allow engineers to compare and cluster functionally similar designs, configurations, codes, geometry, sensor data, etc. This in return will allow engineers to automate the process of quickly searching the required engineering designs available in their library that meet the functional requirements presented in the specification.

9.2   Related Work

Design automation of engineering designs such as electronics, mechanics, etc., has seen an influx in the usage of machine learning and artificial intelligence algorithms [3–7]. These learning algorithms have helped design automation by making the design process easier, faster, and efficient. However, current approaches are mostly limited in the utilization of euclidean domain data and algorithms.

Research on more general non-euclidean domain-based learning algorithms has recently gained momentum [8]. Moreover, a significant amount of work has been done in implementing the convolutional neural network on non-euclidean structure data and manifolds of 3D objects [9–16]. These works can be divided into two general directions in which the learning algorithms have been implemented on non-euclidean data (such as a graph). The first direction is in the spectral domain, and the second direction is in the spatial or vertex domain. In spectral domain-based analysis, just like how filter weight is learned in the traditional 2-dimensional convolutional neural network, filter kernels are learned. In order to do this, first, the graph is transformed into the Fourier domain by projecting the high-dimensional vertex domain graph to low-dimensional space using eigenbasis of the graph Laplacian operation [17]. There are works where the various forms of graph Laplacian operator and approximation methods for reducing the size of the graph kernel and the Fourier transformation of the graphs have been proposed[15, 16, 18–20]. In the second direction, first, the neighborhood information of a vertex is gathered using various techniques. This aggregated neighborhood information is then treated as features and different transformation on these features are proposed [21–25]. The major contribution of this work comes from the fact that the sampling and aggregation can focus on a node's neighborhood, thus not requiring the whole graph to be seen during the sampling and aggregation steps. The sampling and aggregation can be used either by using breadth-first [21] or utilizing both breadth-first and depth-first search [23]. These algorithms can effectively extract the node features based on their neighborhood which can be used to perform clustering and classification. Moreover, it has been shown to be effective in generating the graph embedding using the auto-encoders [24]. Furthermore, some work [20] have even proposed to combine the vertex domain and spectral domain approaches to utilize the strengths of each domain.

Both the spectral domain and vertex domain based approaches have shown tremendous potential in generalizing the graph learning algorithms. However, each of these algorithms has a major weakness. The spectral domain based approach relies highly on the Laplacian matrix. It filters the graph by using the eigenvalues of the Laplacian operator. However, this Laplacian matrix is dependent on the graph, and the filter weights trained in one graph cannot be applied to another one. On the other hand, the vertex domain approaches have shown high efficiency in node-level clustering and classification. Although, there are algorithms such as sub-graph2vec [37], struct2vec [26], a more general CNN based approach for learning rich features from a sub-graph is lacking. In engineering design automation, we would require a more general graph learning algorithm that is able to learn sub-graph or whole-graph property for providing more intuitive functional classification, clustering, or even generation.

In addition, custom graph kernel modeling for mining the graph by measuring the structural similarity between the pairs of graphs has also been proposed [27]. The major limitation of this approach is that they do not consider the features of the individual nodes and only rely on the structural similarity of the graph. Hence, it may be useful in mining structures from design engineering data; however, it will not help if the similar structure has different node features (which normally is the case in the engineering data). To address the existing limitation in the graph learning algorithms, in this chapter we introduce the structural graph convolutional neural network that is graph invariant (unlike spectral domain approaches), can learn rich features from nodes, sub-graph, or the whole graph, and is able to use both the structure and node features to learn rich features to classify and cluster different engineering domain graphs to aid in design automation.

9.3   Graph Learning Using Convolutional Neural Network

Let us define some preliminary notations for explaining the graph structure. The graph is denoted as $G = (V, \mathcal{E})$, where the set of vertices of the graph is denoted as V and the set of edges of the graph is denoted as $\mathcal{E}$. The graph edges can be weighted $w_{ij}, i, j \in V$ and be directed. In this chapter, we will consider unweighted graphs with $w_{ij} = 1$. However, we may easily expand the graph learning algorithm for weighted graphs as well. When the engineering data is represented using such graph structure, each of the vertices will have some features (such as design version, mechanical properties, etc.). We represent such features for each vertices using the symbol f_i, where $i \in V$. f_i consists of a vector whose dimension depends on the amount of information present in the engineering data. The raw format of features may vary from being a text, image, continuous or discrete signals, etc. In such a situation, these features need to be converted to its corresponding vector representation using auto-encoders. The adjacency matrix of the graph G is denoted by $\bar{A}$. The sub-graph of G is denoted as $G_s = (V_s, \mathcal{E}_s)$, where $V_s \subseteq V$ and $\mathcal{E}_s \subseteq \mathcal{E}$.

The proposed graph learning algorithm's architecture is shown in Fig. 9.1. Before the algorithm can be used, the raw engineering design data needs to be converted into a knowledge-based graph [28]. The structure of the knowledge-based graph should be tailored to the engineering domain. The next step will involve converting the high-dimensional information present in each of the vertices and edges to a lower

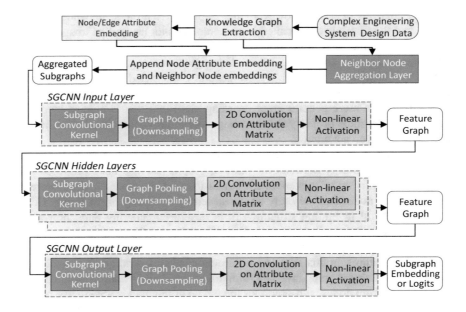

Fig. 9.1 Proposed graph learning algorithm for engineering design automation

dimensional vector space embedding. These embedding will form the features f_i for all the vertices. The major contributions of the proposed graph learning algorithm are (1) neighbor node aggregation layer, (2) sub-graph convolutional kernel, and (3) graph pooling algorithm. However, there are various components of the proposed algorithm which enable it to function. Each of these components is explained in the following subsections.

9.3.1 Knowledge Graph Extraction

The first task for utilizing the proposed graph learning algorithm is converting the complex engineering domain data into a meaningful data structure which can concisely and precisely represent the engineering domain data. For this purpose, we propose to utilize a knowledge graph. The knowledge graph stores information between the various uniquely identifiable entities and their corresponding relationship. These relationships are stored in the form of a triple (node-edge-node) relationship. This type of triples has previously been used to create knowledge graphs such as DBPedia [29]. The main advantage of using such knowledge graphs is that it can store rich engineering domain information and continuously evolve, grow, and be linked to other engineering domain data as well.

9.3.2 Attribute Embedding

After the knowledge graph has been created, we will have the structure of the graph ready to be utilized in the graph learning algorithm. However, in a knowledge graph, the node and edge may be in different data formats (such as text, images, etc.). These high-dimensional attribute of the nodes and edges needs to be converted into a low-dimensional feature embedding. To embed such attributes we utilize various state-of-the-art auto-encoders. For example, for encoding the text we will utilize word2vec [30]. For embedding images, we will utilize existing deep auto-encoder [31]. These vector embedding generated from the attribute of the nodes and edges form the feature which is utilized in the attribute matrix described in the Sect. 9.3.4.1.

9.3.3 Neighbor Nodes Aggregation

One of the fundamental tasks in performing graph learning in engineering domain data is being able to capture the features of the vertices with respect to its location in the graph. Each of the vertices not only has special topology but also shares a set of attributes across the knowledge graph. Hence, it is necessary to capture each of the unique structural and feature based relation of vertices with respect to its neighbor. In order to do this, in the proposed graph learning algorithm, we utilize concepts similar to the vertex domain approach [21] (see Fig. 9.2). Before the neighbor node aggregation is performed, a user-defined query or a *schema* is used to induce a sub-graph. This induction process is domain-specific and can further be tailored to aid the graph learning algorithm by inducing sub-graphs that capture the meaning of engineering design. For example, the schema can be used to induce graphs that contain certain keywords (such as engines, valves, etc.,) if the graph learning is applied to design data of an engine. Based on this schema various instances are generated by the sample generator. These induced samples are then passed to two blocks in parallel. One block converts the node/edges attribute to its corresponding vector form, whereas the other block is the neighbor node aggregation layer. In the neighbor node aggregation layer, for the induced sub-graph $G_t = (V_t, \mathcal{E}_t)$, our algorithm performs both breadth-first and depth-first search to collect the neighbor nodes of the given sub-graph. We utilize the parameter d as the depth to search in the graph, and n as the number of paths to be searched. The neighbor nodes borrow the concept from node2vec [23], where a balance is maintained between the depth and the breadth for context generation. This balance helps us to maintain a balance between the local and non-local neighbors to the sub-graph.

In order to find the neighbors, for all $v_i \in V_t$, we search the original knowledge graph G to find n paths of length d. Let us denote each of these paths by $\mathcal{P}_i^d$, where d denotes the path length, and i denoted the ith path of length d. For the path to be considered for aggregation the nodes lying in the path $v_i \notin V_t$. We acquire vector

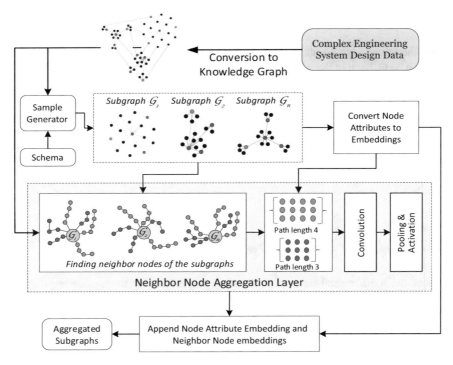

Fig. 9.2 Proposed neighbor node aggregation layer architecture

of paths $\mathcal{P}^d = \{\mathcal{P}_1^d \, \mathcal{P}_2^d, \ldots\}$ for each of the sub-graph $\mathcal{G}_t$. However, finding and using all of such paths is non-trivial, hence we then sample s paths of length d for each of the sub-graph. This sampling is done randomly in the proposed algorithm. Hence, s is another input parameter to the proposed graph learning algorithm. Using these randomly sampled paths, we form a neighbor feature matrix $\bar{N}$. The bar is s by d matrix with each element having a feature vector having extracted earlier. The number of paths found in $\mathcal{P}^d$ may be smaller than s, $\mathcal{P}^d$ can be padded to make $\bar{N}$ with at least s number of rows/paths. The algorithm for neighbor node aggregation is presented in Algorithm 1.

The input to the Algorithm 1 is the knowledge graph $\{\mathcal{G}_1, \mathcal{G}_2, \ldots, \mathcal{G}_t\}$ using the user-defined schema, list of depth to search $D = \{d_1, d_2, \ldots, d_n\}$, list of sample numbers per depth $S = \{s_1, s_2, \ldots, s_n\}$, and feature vector x_n for the vertices. The output of the Algorithm 1 is the extracted feature to be appended to the vertices of the sub-graph G_t. One of the challenges in extracting the features from the neighborhood nodes as discussed in [21] is the fact that the non-euclidean data has no natural ordering, which means that the feature extraction should be applied over an unordered set of paths to make sure that the arbitrary change in the order of rows of the matrix $\bar{N}$ still results in the same feature being extracted. In order to achieve this in Algorithm 1, we first apply a general 1-D convolution operation

---

Algorithm 1: Neighbor node aggregation algorithm

Input: Induced Sub-graphs: $\mathcal{G}_1, \mathcal{G}_2, \ldots, \mathcal{G}_t$
Input: A list of depth to search: $D = d_1, d_2, \ldots, d_n$
Input: A list of sample numbers per depth: $S = s_1, s_2, \ldots, s_n$
Output: A feature vector: x_n

1 **foreach** $d_i \in D$ **do**
2   **foreach** $v_j \in \mathcal{V}_t$ **do**
3   Find all length d_i paths $P_j^{d_i}$
4   Remove paths containing nodes in $\mathcal{V}_t$ from $P_j^{d_i}$
5   Add $P_j^{d_i}$ into P^{d_i}
6   Construct $\bar{N}$ by randomly selecting s_i paths from P^{d_i}
7   Extract feature x_{di} from $\bar{N}$
8 $x_n = f_{pool}^d(x_{d1}, x_{d1}, \ldots, x_{dn})$
9 **return** x_n

---

with trainable 1 by d weight matrix $\bar{W}$ on $\bar{N}$ and then utilize a symmetric pooling function to extract the neighbor nodes' feature vector x_n as follows:

$$x_n = \sigma(f_{pool}(\bar{W} \circledast \bar{N}) + b) \tag{9.1}$$

This equation is used in Line 8 of the Algorithm 1. The b in Eq. 9.1 is a bias variable and σ is an activation function (e.g., ReLU, LeakyRelu, Sigmoid, Tanh, etc.), and f_{pool} is a pooling function. As mentioned, the pooling function has to be invariant to permutation of rows in $\bar{N}$. To achieve this, pooling function such as a mean operator can be applied over all the rows in the matrix, or we may utilize a max pool operator that extracts max values out of all the rows. The use of a specific pooling function is treated as a hyper-parameter in the graph learning algorithm and later configured in the training and hyper-parameter tuning process. As shown in Fig. 9.2, we extract various path lengths $\{d_1, d_2, d_3, \ldots, d_k\}$. These path lengths will integrate various topological and localized attributes of the engineering knowledge graph. Hence, Algorithm 1 will return feature vector $\{x_{d1}, x_{d2}, x_{d3}, \ldots, x_{dk}\}$ depending on total number of path lengths extracted from the graph. Each of these path lengths is aggregated to the sub-graph. If the extracted path length number is large, we may perform pooling to reduce the dimension of the extracted features as:

$$x_n = f_{pool}^d(\{x_{d1}, x_{d2}, x_{d3}, \ldots, x_{dk}\}) \tag{9.2}$$

The returned neighborhood feature vector is then concatenated to all the feature vectors of the vertices $v_i \in \mathcal{V}_t$ as $x_{agg_i} = \{x_i, x_n\}$. Since the neighborhood aggregation layer is part of the graph learning algorithm, all the parameters of the neighborhood aggregation layer (such as weights of the 1-D convolution neural network) are learned during training. This allows the algorithm to automatically focus on relevant neighborhood node features to aggregate during the training.

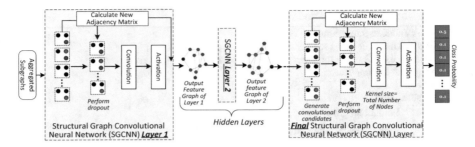

Fig. 9.3 Proposed structural graph convolutional neural network architecture

9.3.4 Structural Graph Convolutional Neural Network Layers

The structural graph convolutional neural network (SGCNN) layers are shown in Fig. 9.3. The input of the SGCNN are the aggregated sub-graphs that were generated by the neighbor node aggregation layer described in Sect. 9.3.3 earlier. Each of these aggregated sub-graphs is passed in batches with individual vertices having a feature matrix. The SGCNN layers consist of input, hidden, and output layers. These layers abstract the aggregated graph's feature in each layer just like the traditional 2D-convolutional neural networks. Each of the SGCNN consists of various components: (1) sub-graph convolution kernel, (2) graph pooling, (3) 2D convolution on adjacency matrix, (4) new adjacency matrix calculation, and (5) non-linear activation. We will discuss each of these components in detail in the following sections.

9.3.4.1 Sub-Graph Convolution Kernel

The main block of the SGCNN is the sub-graph convolutional kernel. The task of the sub-graph convolutional kernel is to abstract meaningful feature vectors from the aggregated graph. The kernel receives the aggregated graph G_t with the corresponding feature matrix $\bar{X}$, which consists of the aggregated features $x_{agg_i} = \{x_i, x_n\}$ for $v_i \in \mathcal{V}_t$. It also receives the adjacency matrix $\bar{A}$ of the aggregated graph G_t. With the adjacency matrix and the feature matrix, the first step the sub-graph convolutional kernel block will do is create the adjacency matrix $\bar{A}r$ by taking the Hadamard product between $\bar{X}$ and $\bar{A} + \bar{I}$ as follows:

$$\bar{A}r = \bar{X} \circ (\bar{A} + \bar{I}) \tag{9.3}$$

We have added the identity matrix $\bar{I}$ to make sure that the vertices do not lose their feature information. This self-loop to each of the vertices in the sub-graph makes sure that each vertex retain their own information while calculating the Hadamard product. An example of an attribute matrix is shown in Fig. 9.4. The attribute matrix

$$
\begin{bmatrix}
x_1 \ x_2 \dots x_{n-1} \ x_n \\
x_1 \ x_2 \dots x_{n-1} \ x_n \\
\vdots \quad \vdots \quad \vdots \quad \vdots \\
x_1 \ x_2 \dots x_{n-1} \ x_n \\
x_1 \ x_2 \dots x_{n-1} \ x_n
\end{bmatrix}
\circ
\begin{bmatrix}
1 \ 1 \dots 1 \ 0 \\
1 \ 1 \dots 0 \ 0 \\
\vdots \quad \vdots \quad \vdots \quad \vdots \\
1 \ 1 \dots 1 \ 0 \\
0 \ 0 \dots 0 \ 1
\end{bmatrix}
=
\begin{bmatrix}
x_1 \ x_2 \dots x_{n-1} \ 0 \\
x_1 \ x_2 \dots 0 \ 0 \\
\vdots \quad \vdots \quad \vdots \quad \vdots \\
x_1 \ x_2 \dots x_{n-1} \ 0 \\
0 \ 0 \dots 0 \ x_n
\end{bmatrix}
$$

Fig. 9.4 Example of an attribute matrix calculated using Hadamard product

$\bar{A}r$ is used to define a graph convolutional kernel. The kernel consists of a k by k weight matrix $\bar{W}^k$. The size of the k can be varied just like the 2D-convolutional kernels. The convolutional is applied between the $\bar{W}^k$ and $\bar{A}r$. In traditional 2D convolution, a kernel is slid from left to right and top-down in an order to perform the 2D convolution. However, since there is no notion of 2D-grid structure in the graph, we cannot perform convolution in a similar manner. Hence, in our graph convolution, we propose a new method to perform a convolution operation for the graph data structure.

The algorithm for performing the graph convolution is presented in Algorithm 2. The first step in performing the graph convolution involves generating candidate graph kernels with size k by k, where we select the k vertices at a time. This candidate graph is then convoluted with the weight matrix $\bar{W}^k$. In order to generate the candidate graph kernels, we use the fact that removing the ith row and ith column in $\bar{A}r$ is equivalent to removing the vertex i from $\mathcal{G}_t$. Hence, assuming that the total vertices in the induced aggregated sub-graph $\mathcal{G}_t$, let it be denoted by n, are greater than the size of the kernel k, we will remove $n - k$ number of vertex from $\mathcal{G}_t$. The new sub-graph will be left with a new k by k attribute matrix $\bar{A}r^k$. The drawback of generating the attribute matrix like this is that there are $O\binom{n}{k}$ possible ways to generate the attribute matrix $\bar{A}r^k$. This might be okay for an induced graph with a lower number of nodes; however, for engineering data the induced graph size can have a large number of nodes. And normally the size of the $k \ll n$, making the complexity of generating the $\bar{A}r^k$ very high. Hence, to tackle this impractical $\bar{A}r^k$ generation step, we propose to relax it by only picking s number out of $O\binom{n}{k}$ as a convolution candidate. Hence, the total number of the possible $\bar{A}r^k$ is thus reduced to $O(s)$. By doing this, we make it feasible to generate the $\bar{A}r^k$ from the $\mathcal{G}_t$ as a potential candidate of graph kernels to be convoluted with the k by k weight filter matrix $\bar{W}^k$. The procedure of down-sampling of $O\binom{n}{k}$ possible $\bar{A}r^k$ values to just s is explained in Sect. 9.3.4.2).

Algorithm 2 presents the graph convolution steps in detail. The input to the graph is the induced aggregated graph $\mathcal{G}_t$. The output of the algorithm is the new graph where each node represents the merged vertices present in $\bar{A}r^k$. In Line 1, we first generate the adjacency matrix from the graph. In Line 4, the attribute matrix is generated by performing the Hadamard product between the feature matrix and the adjacency matrix. In Line 7, possible combination of $\bar{A}r^k$ is listed and down-

---

Algorithm 2: Graph convolution calculation algorithm

Input: An input graph: G_t with n vertices
Input: A convolution kernel: $\bar{W}^k$
Input: Sample size: s
Output: An output feature graph: G'
1 Generate adjacency matrix $\bar{A}$ from G
2 Using the same vertices order to generate list of features X
3 Create feature matrix $\bar{X}$ with n rows, and each row being X
4 $\bar{A}r = \bar{X} \circ (\bar{A} + \bar{I})$
5 $m = \binom{n}{k}$
6 $CombList$ = Enumerating choice of $n - k$ elements from $1, 2, \ldots, n$
7 **foreach** $comb \in CombList$ **do**
8 $\quad$ $\bar{A}r_{comb}$ = remove rows and columns list in $cand$ from $\bar{A}r$
9 $\quad$ Add $\bar{A}r_{comb}$ into $CandList$
10 **if** $m > s$ **then**
11 $\quad$ Down-sample $CandList$ to s elements
12 **else if** $m < s$ **then**
13 $\quad$ Pad $CandList$ to s elements
14 **foreach** $cand \in CandList$ **do**
15 $\quad$ $x^k = \bar{W}^k \circledast cand + b$
16 $\quad$ Add new vertex v_k into G'
17 $\quad$ Add feature vector x^k on v_k
18 $\quad$ Connect v_k based on $cand$'s connection in G
19 **return** G'

---

sampled in Line 10. In Line 14, a 2D convolution is performed between the $\bar{A}r^k$ and the filter weight matrix W^k. Finally, the new graph is returned in Line 19. Each of the SGCNN layers will generate a new graph which progressively abstracts and fuses the topological and attributes of the previous graph.

9.3.4.2 Graph Pooling Algorithm

When we generate $\bar{A}r^k$, there are $O\binom{n}{k}$ possible in the beginning. After down-sampling it to s number of $\bar{A}r^k$, the next stage will have $O\binom{s}{k}$ possible combination of $\bar{A}r^k$. One of the desired properties of convolutional neural networks is to be able to perform the convolution over a deep number of layers. Hence, to be able to perform the graph convolution in deeper layers, we need to perform down-sampling or pooling at each layer to manage the size of the possible $\bar{A}r^k$ matrices generated from the graph. Without down-sampling, it will be unfeasible to perform a large number of convolutional operation between $\bar{A}r^k$ and W^k. Hence, we propose to perform the pooling operation before the convolution operation in each of the SGCNN layers. The proposed down-sampling/pooling operation utilizes the topology of the graph to remove samples from combinations of $\bar{A}r^k$. For each of the possible $\bar{A}r^k$, we calculate the corresponding total degrees. The intuition behind the proposed down-sampling algorithm is that out of $O\binom{n}{k}$, due to the sparsity of the aggregated sub-graph, combination of various nodes will not have any edges or lower number of edges among them. Hence, combining them together to perform convolution will be less fruitful than selecting the combination of $\bar{A}r^k$ that have higher connectivity among the vertices. The proposed down-sampling algorithm is presented in Algorithm 3.

Algorithm 3: Graph pooling algorithm

Input: An input graph: G
Input: The list of the candidate nodes combinations: $Comb$
Input: Sample size: s
Input: Dropout Rate: d
Output: The list of down-sampled candidate nodes: $Comb'$

1 Randomly sample $Comb'$ from $Comb$ using d

2 Generate adjacency matrix $\bar{A}$ from G for only $Comb'$

3 **foreach** $c \in Comb'$ **do**

4    $d_c = 0$

5   **foreach** $n \in c$ **do**

6   Calculate Degree of n as d_n

7       $d_c = d_c + d_n$

8   **foreach** $c' \in Comb'$ **do**

9   **if** c' *is connected with* c *in* $\bar{A}$ **then**

10          $d_c = d_c + 1$

11 Keep the s number of nodes with the highest degrees and store in $Comb''$

12 **return** $Comb''$;

The input to the Algorithm 3 is the graph G, the possible list of candidate combination $Comb$, the pooling sample size s, and the dropout rate d. The output of the algorithm is the list of combination $Comb''$ which is down-sampled. To achieve this, first in Line 1, it randomly samples $Comb'$ by using the dropout rate d. This is necessary because it is computationally unfeasible to calculate the adjacency matrix in the next step for all the possible candidates in $Comb$. In Line 2, we generate the adjacency matrix $\bar{A}$ for the new candidate combination of nodes. This adjacency matrix carries the graph structural data and passes it through the deeper layers. This step is important, as the different layers will be able to abstract the graph data in a hierarchical manner. Lines 3–10 compute the total degrees of each candidate nodes combination $Comb'$, which are combinations of the nodes in G that are generated by the convolution kernel and these combinations will serve as the new nodes of the output feature graph after the graph convolution kernel. Specifically, Lines 3–7 compute the total degrees inside the combination, and Lines 8–10 compute the degrees in between different combinations. Finally, we keep the s number of nodes combinations which have the highest degree and remove the rest. We significantly reduce the size of the graph convolution kernel by dropping the combinations in the calculated degrees using the max pooling. Nevertheless, we ensure that the convolution is performed on the graph structures with higher connectivity.

9.3.4.3   2D Convolutions on Attribute Matrix

After we have down-sample s possible candidate $\bar{A}r^k$ matrices, we apply simple 2D convolution operation to extract the feature vector for the given combination of the $\bar{A}r^k$. The convolution operation can be written as follows:

$$x^k = \phi(\bar{W}^k \circledast \bar{A}r^k + b) \tag{9.4}$$

where $\phi(.)$ is a non-linear activation function. We will have s extracted feature vectors as: $x_1^k, x_2^k, \ldots, x_s^k$. We consider the extracted feature vectors $x_1^k, x_2^k, \ldots, x_s^k$ as a new feature graph $\mathcal{G}'$ with s number of vertices, and x_i^k as the feature vector for node i. An example of this process is shown in Fig. 9.5. Given the input graph with 4 vertices, a 3-by-3 convolution kernel is selected. As a result, 4 convolution candidates are generated, down-sampled, and adjacency recalculated resulting in a new graph with 3 vertices.

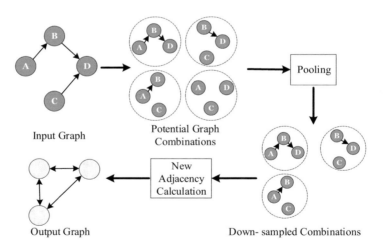

Fig. 9.5 Convolution kernel example

9.3.4.4 New Adjacency Matrix Calculation

For the deep SGCNN architecture to work, we have to keep track of the adjacency of the new G' generate at each of the layers. This G' is used as input to another sub-graph convolution Layer to form a deep SGCNN model. The constructed adjacency matrix will allow the next SGCNN layer to recalculate the new attribute matrix and the potential candidates of $\bar{A}r^k$ to be selected for convolution and down-sampling. To calculate the new adjacency for the new graph, we check the edges between inter and intra nodes of the graph convolution kernels. The edge between these intra and inter nodes allows the graph structure to be propagated through the SGCNN layers, making sure the topological information is being utilized to learn the filter weight matrix W^k at each layer.

9.3.5 Classification for Engineering Design Abstraction

Given a large graph and labeled sub-graphs representing engineering design, the SGCNN can be used to classify them. While classifying these sub-graphs, various features regarding the design are learned by the graph convolutional kernels. Once the design and the labels (which can be the function served by the design) are trained, the design is abstracted based on their functions. This classification is done by using a softmax function and cross-entropy of the logits. In addition, the feature vectors (sub-graph embedding) generated by the final SGCNN can also be used by clustering algorithms to identify nearest neighbors sub-graphs that have an equivalent function in the graph based on their node attributes and structure. In engineering, there are several use-cases for sub-graph embedding including the identification of functionally equivalent structures that engineers are unaware of, and to identify structures that mislabeled.

9.3.6 Graph Learning Algorithm Hyper-Parameters

In deep learning convolutional neural networks, hyper-parameters play a crucial role in improving the accuracy and convergence of the algorithm. These hyper-parameters are difficult to derive during the training as it requires large resources. Hence, to ease the resources they are instead selected prior to training. Grid search or other efficient approaches are utilized to fine-tune the hyper-parameter values. For euclidean domain, extensive study [32, 33] have been carried out to aid in the hyper-parameter selection. Some of the hyper-parameters that even common to non-euclidean deep graph convolutional neural networks are *activation function, hidden layers, number of iteration, learning rate, and batch size*. However, as the proposed deep graph learning algorithm for the engineering design automation is relatively new, there are few other hyper-parameters that need to be highlighted. These hyper-parameters are as follows:

9.3.6.1   Path Length in Node Aggregation Layer

As mentioned in Sect. 9.3.3, the neighbor node aggregation layer utilizes a specific path number of various length sizes. The path length determines how much of the knowledge graph should be considered in embedding the feature for the given induced sub-graph. If the length is small, then we will focus on the local structure and attributes of the sub-graph, while selecting longer path length will allow the sub-graph to be embedded with global features. Hence, this value needs to be optimized and fine-tuned according to the specific engineering domain data.

9.3.6.2   Graph Convolution Kernel Size

The graph convolutional kernel size determines how many vertices to be considered for convolution with the filter weight matrix at each layer. The smaller kernel size means that each of the layers will abstract sub-graph feature by considering its immediate neighbors. However, if the kernel size is small while the number of vertices in the induced sub-graph is large, then deeper SGCNN layers may be needed. However, if we select larger kernel the SGCNN layer may be shallow. The size of the kernel may depend on the induced graph's size and structure and will require tuning before the training is performed.

9.3.6.3   Dropout of Candidate Kernels

As presented in Sect. 9.3.4.2, we have used dropout to make the deep graph learning feasible. We have combined the random and degree-based dropout. If permitted by the resource, taking a large number of candidate attribute matrix may be helpful to better abstract the induced sub-graph. Due to the sparse nature of the engineering design data taking large combinations of attribute matrix for convolution may also be a waste of resources. Hence, a number of candidate kernels to drop out before the convolution is performed need to be tuned as a hyper-parameter.

9.4   GrabCAD Dataset

To demonstrate the applicability of the proposed graph learning algorithm in engineering design automation, we have selected the 3D engineering CAD models as training as a testing dataset for engineering design functionality classification.

Fig. 9.6 Sample of the 3D CAD models extracted from GrabCAD repository

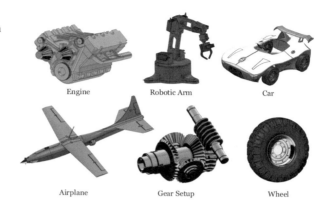

Engine   Robotic Arm   Car

Airplane   Gear Setup   Wheel

We have extracted the dataset from GrabCAD,[1] which is an online repository of 3D CAD models shared and maintained by an online community of designers, engineers, and manufacturers. It consists of over four million members with over two million engineering design models. From this vast online dataset, we have extracted meta-information from six functional categories of 3D CAD models. These functional category are *Car*, *Engine*, *Robotic arm*, *Airplane*, *Gear*, and *Wheel* (Fig. 9.6).

Since these are engineering designs from the mechanical domain, we have tailored a custom schema consisting of the properties such as 3D model's name, author of the design, description of the model, parts names, tags, likes, time-stamps, and comments on the engineering design. With this schema we have induced sub-graphs for each of the categories with 2271 samples for *Car*, 1597 samples for *Engine*, 2013 samples for *Robotic arm*, 2114 samples for *Airplane*, 1732 samples for *Gear*, and 2404 samples for *Wheel*. The induced sub-graph consists of 17 nodes consisting of both social network data (such as user-to-user interaction through comments and likes) and engineering data (such as model-to-tags relationship, model-to-model relationships, etc.). By inducing a sub-graph in GrabCAD dataset, we aim to demonstrate that similar designs with functional description represented by a knowledge graph can be efficiently classified using the proposed supervised graph learning algorithm.

9.5 Results

In our experiment, the total number of the induced graph from all the engineering design is 14,131. From this sample, we have used 11,304 as training samples and 2827 samples for testing if the similar functional designs get classified accurately.

[1] https://grabcad.com/.

In order to make sure that proper hyper-parameters are selected and tuned, we have used a grid search approach over the possible hyper-parameter values. The result of hyper-parameter selection is shown in Table 9.1. In the table, we have tested the accuracy of various hyper-parameters. The first hyper-parameter is the learning rate of the optimization algorithm. We have used ADAM optimizer [34] to adjust the filter weights. From the table, it can be seen that the learning rate of 0.01 is able to achieve higher accuracy. The second hyper-parameter in the table is the batch size. The batch size determines how many of the induced sub-graph are passed together once for calculating the loss and updating the gradient.

It can be seen the batch size of 32 is able to obtain the highest accuracy of 79.22% classification accuracy. The next hyper-parameter is the hidden layer feature size. Each of the SGCNN layers is able to determine the size to give as an output. It can be noticed that the higher feature size of 150 is able to achieve better accuracy. With a larger feature size, we also increase the number of parameters in the filter weight matrix. The nest hyper-parameter is the epoch (the total number of times the training goes over the whole dataset). It can be seen that a higher epoch number of 500 is able to achieve higher classification accuracy. The last hyper-parameter is shown in Table 9.1 is the output layer kernel size. The final output layer of the SGCNN acts like a dense layer, which means that it will try to generate probability values for each of the categories. The final layer of the kernel depends on the dropout carried out earlier. From the table, it can be seen that lower dropout or larger output kernel produces higher accuracy. Beside these hyper-parameters, we have also tested other hyper-parameters which are presented in the following sections.

Table 9.1 Accuracy for various hyper-parameters

Learning rate	1.00E-04	1.00E-03	1.00E-02	1.00E-01	Hidden layer feature = 100, Epoch = 100, Output layer kernal size = 25, Batch size = 64,
Accuracy (%)	45.70	66.051	79.26	80.15	
Batch size	32	64	128	256	Hidden layer feature = 100, Epoch = 100, Learning rate = 1e-2 Output layer kernel size = 25,
Accuracy (%)	79.22	78.65	76.7	72.01	
Hidden layer feature size	20	50	100	150	Output layer kernel size = 25, Epoch = 100, Learning rate = 1e-2 Batch size = 64,
Accuracy (%)	82.28	83.10	85.54	87.89	
Epochs	100	200	300	500	Hidden layer feature = 100, batch = 64, Learning rate = 1e-2 Output layer kernal size = 25,
Accuracy (%)	78.87	79.68	80.43	81.12	
Output layer kernel	25	30	40	45	Hidden layer feature = 100, Epoch = 100, Learning rate = 1e-2 Batch size = 64,
Accuracy (%)	85.61	85.22	86.43	87.92	

Three Layers (Aggregate, SGCNN Input, SGCNN Output), Random Dropout

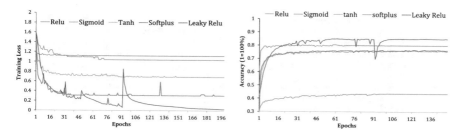

Fig. 9.7 Training loss and accuracy for different activation functions

9.5.1 Activation Functions

Activation functions are used before the output in each of the SGCNN layers. Most of the operation before the activation function is mostly linear. However, the activation function increases the capacity of the SGCNN layer by making it non-linear. We have explored various activation functions which are well studied in euclidean domain. These activation functions are *sigmoid, softplus, tanh, rectifier linear unit*, and *leaky rectifier linear unit*. The training loss and engineering design classification accuracy during testing are shown in Fig. 9.7. The configuration for acquiring the results consists of layers size of 2 consisting of aggregate and embedding layer only. It can be seen that out of all activation functions, leaky rectifier linear unit ($f(x) = \alpha x$ for $x < 0$ and $f(x) = x$ for $x >= 0$) activation function (with $\alpha = 0.2$) is able to achieve lower loss and higher classification accuracy compared to other activation functions. Although the accuracy is higher, it introduces some noise in both the loss and the accuracy values. One of the parameters that can be used in *LeakyRelu* is the alpha value. It determines the slope to be used to cut off the values when $x < 0$. Further analysis is required to tune the value of the α.

9.5.2 Kernel Size

One of the important hyper-parameters for the SGCNN layers is the size of the kernel used to perform the convolution with the attribute matrix $\bar{A}r^k$. The value of k determines the total number of vertices that are considered at a time to perform the convolution with the filter weights W^k. In our experiment, the total of nodes available in the induced sub-graph is 17, hence we have selected kernel size as 2, 4, 6, 8, 10, 12, and 14. It can be seen from Fig. 9.8 that the kernel size has a drastic effect on the classification accuracy of the engineering design data. The configuration of SGCNN consists of random dropout, last layer kernel size of 5, three total layers, relu as hidden layer activation, and leaky relu as the final layer

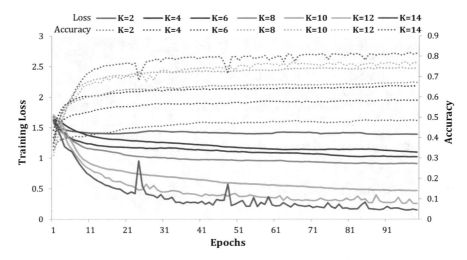

Fig. 9.8 Training loss and accuracy for different sizes of kernels

activation function. For three layers of SGCNN, larger kernel size is able to achieve higher testing accuracy and lower training loss. However, having a larger kernel size in every hidden layer is not feasible, as it increases the complexity of the training algorithm. In Sect. 9.5.4, it can be seen that with deeper layers, a smaller kernel size is able to obtain higher testing classification accuracy as well.

9.5.3   Dropout

As mentioned earlier, without dropout, it becomes unfeasible to create deep SGCNN layers. To improve the scalability of the SGCNN layers, in the proposed graph learning algorithm we utilized a combination of random and adjacency based dropout. In the adjacency based dropout, we utilize the degree of the vertices to select the candidate combination of $\bar{A}r^k$ for the convolution. The dropout rate determines the size of the final kernel size. If the dropout rate is higher, the size of the kernel at the output SGCNN layer will be smaller and vice versa. The result of the various size of the final layer's kernel size as a result of changing the dropout is shown in Fig. 9.9. The SGCNN configuration consists of hidden layer kernel size of fourteen, total of three layers, relu as hidden layer activation, and leaky relu as the final layer activation function. It may be noticed that if the dropout is less (resulting in larger kernel size in the output layer), the testing accuracy is higher and training loss is lower. The result is shown for just three layers of the SGCNN. The total number of possible combination of $\bar{A}r^k$ with the kernel size of 14 for the first layer is $\binom{17}{14} = 680$. We can notice that even when the total candidates have been drastically

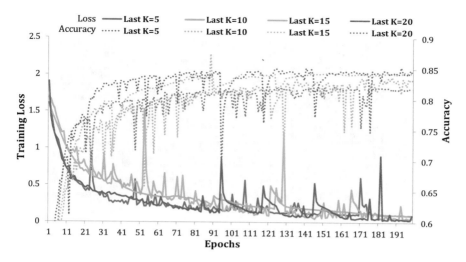

Fig. 9.9 Training loss and accuracy for various random dropouts

dropped to just 5, 10, 15, and 20, the graph learning algorithm is able to perform quite well in classifying the engineering designs. This may be due to the fact that the induced graphs are sparse in nature and that the initial kernel size of 14 is able to capture all of the node's features during convolution.

In addition to random dropout, we have implemented the adjacency based dropout as well. The down-sampling algorithm first uses random dropout to initially reduce the possible candidates of $\bar{A}r^k$, and out of the remaining selects the ones with higher connectivity. The result for the down-sampling combined with the random dropout is shown in Fig. 9.10. It may be noticed that adjacency based dropout is able to achieve the highest accuracy of around 90%.

9.5.4 Layers

The most advantageous property of the proposed graph learning algorithm is able to have deeper layers that are able to abstract the features of the induced graph in each iteration. To demonstrate this capability, we have selected a kernel size of 2 for the filter weights, and measure the accuracy of the graph learning algorithm for layer size of 3, 4, 5, and 6. Furthermore, the hidden layer kernel size is set to two, leaky relu is chosen as the activation function, last layer kernel size is set to thirty, and random dropout is selected. It can be noticed from Fig. 9.11 that layer size of 4 and 5 is able to achieve higher testing classification accuracy compared to shallow 3 layers and deeper 6 layer size. The highest accuracy achieved was ≈91% with four layers. The deep SGCNN layers are able to abstract the structural and

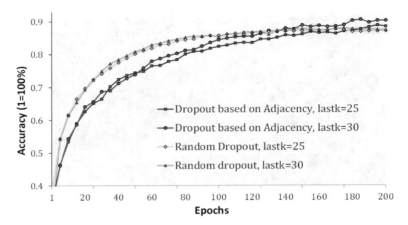

Fig. 9.10 Comparison between random dropout and adjacency based dropout

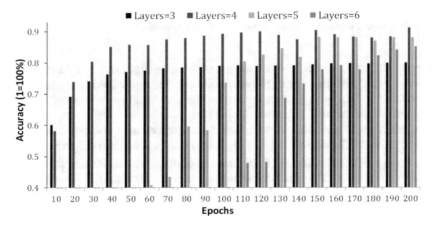

Fig. 9.11 Performance comparison between various layer sizes

attribute properties of the induced sub-graph in a hierarchical manner by using a smaller kernel size of 2. This means that in each layer smaller node size is fused and the new combined features are learned. This property can be helpful in engineering design data, where there is some form of hierarchy in terms of design.

9.6   Discussion

The SGCNN's capability to learn sub-graph structure embedded with attributes was demonstrated in Sect. 9.5. It achieves very positive results on functional lifting using the GrabCAD dataset. We have presented the GrabCAD dataset as an engineering dataset to show the applicability of design automation. Although SGCNN was created to address the functional lifting problem in engineering, we believe this is broadly applicable to other domains. Currently, we have shown that SGCNN is able to perform supervised learning in abstracting the design features and classifying them based on their functions. For robust design automation, designers would be interested not only in classifying the designs but being able to reproduce or generate designs that slightly vary in functionality. In order to do this, the same structure of the SGCNN may be used for generative learning algorithms such as variational auto-encoder (VAE) [35] or generative adversarial networks (GAN) [36].

9.7 Summary

In this chapter, we presented a novel structural graph convolutional neural network which can be used to abstract the non-euclidean graph or sub-graph dataset. These graphs can be used to represent the engineering design data and allow designers to effectively perform design automation by effectively clustering engineering designs with similar functionality. This allows designers to search for designs that are similar to the required functionality and aid in design automation.

References

1. Haik, Y., Sivaloganathan, S., & Shahin, T. M. (2018). *Engineering design process*. Scarborough: Nelson Education.
2. Wan, J., Pollard, B. S., Chhetri, S. R., Goyal, P., Faruque, M. A. A., & Canedo, A. (2018). Future automation engineering using structural graph convolutional neural networks. Preprint arXiv:1808.08213.
3. Li, B., & Franzon, P. D. (2016). Machine learning in physical design. In *2016 IEEE 25th Conference on Electrical Performance Of Electronic Packaging And Systems (EPEPS)* (pp. 147–150). Piscataway: IEEE.
4. Pandey, M. (2018). Machine learning and systems for building the next generation of EDA tools. In *Proceedings of the 23rd Asia and South Pacific Design Automation Conference* (pp. 411–415). Piscataway: IEEE.
5. Qi, W. (2017). *IC design analysis, optimization and reuse via machine learning*. Raleigh: North Carolina State University.
6. Kazi, R. H., Grossman, T., Cheong, H., Hashemi, A., & Fitzmaurice, G. W. (2017). Dreamsketch: Early stage 3D design explorations with sketching and generative design. In *UIST 2017 Conference proceedings: ACM Symposium on User Interface Software & Technology* (pp. 401–414).
7. Chen, X. A., Tao, Y., Wang, G., Kang, R., Grossman, T., Coros, S., et al. (2018). Forte: User-driven generative design. In *Proceedings of the 2018 CHI Conference on Human Factors in Computing Systems* (p. 496). New York: ACM.
8. Geometric Deep Learning (2018). http://geometricdeeplearning.com/
9. Tenenbaum, J. B., Silva, V. D., & Langford, J. C. (2000). A global geometric framework for nonlinear dimensionality reduction. *Science, 290*(5500), 2319–2323 (2000).

10. Roweis, S. T., & Saul, L. K. (2000). Nonlinear dimensionality reduction by locally linear embedding. *Science, 290*(5500), 2323–2326 (2000).
11. Belkin, M., & Niyogi, P. (2003). Laplacian eigenmaps for dimensionality reduction and data representation. *Neural Computation, 15*(6), 1373–1396 (2003).
12. Hadsell, R., Chopra, S., & LeCun, Y. (2006). Dimensionality reduction by learning an invariant mapping. In *2006 IEEE Computer Society Conference on Computer Vision and Pattern Recognition, 2* (pp. 1735–1742). Piscataway: IEEE.
13. Cao, S., Lu, W., & Xu, Q. (2015). Grarep: Learning graph representations with global structural information. In *Proceedings of the 24th ACM International on Conference on Information and Knowledge Management* (pp. 891–900). New York: ACM.
14. Henaff, M., Bruna, J., & LeCun, Y. (2015). *Deep convolutional networks on graph-structured data*. Preprint. arXiv:1506.05163.
15. Defferrard, M., Bresson, X., & Vandergheynst, P. (2016). Convolutional neural networks on graphs with fast localized spectral filtering. In *Advances in Neural Information Processing Systems* (pp. 3844–3852).
16. Kipf, T. N., & Welling, M. (2016). *Semi-supervised classification with graph convolutional networks*. Preprint. arXiv:1609.02907.
17. Chung, F. R. K. (1997). *Spectral graph theory*, Number 92. Providence: American Mathematical Society.
18. Levie, R., Monti, F., Bresson, X., & Bronstein, M. M. (2017). *Cayleynets: Graph convolutional neural networks with complex rational spectral filters*. Preprint. arXiv:1705.07664.
19. Yi, L., Su, H., Guo, X., & Guibas, L. J. (2017). SyncSpecCNN: Synchronized Spectral CNN for 3D shape segmentation. In *Computer Vision and Pattern Recognition* (pp. 6584–6592).
20. Abu-El-Haija, S., Kapoor, A., Perozzi, B., & Lee, J. (2018). *N-GCN: Multi-scale graph convolution for semi-supervised node classification*. Preprint. arXiv:1802.08888.
21. Hamilton, W., Ying, Z., Leskovec, J. (2017). Inductive representation learning on large graphs. In *Advances in Neural Information Processing Systems* (pp. 1025–1035). New York: ACM
22. Shervashidze, N., Schweitzer, P., van Leeuwen, E. J., Mehlhorn, K., & Borgwardt, K. M. (2011). Weisfeiler-Lehman graph kernels. *Journal of Machine Learning Research, 12*, 2539–2561.
23. Grover, A., & Leskovec, J. (2016). Node2vec: Scalable feature learning for networks. In *Proceedings of the 22nd ACM SIGKDD International Conference on Knowledge Discovery and Data Mining* (pp. 855–864). New York: ACM.
24. Goyal, P., Ferrara, E. (2018). Graph embedding techniques, applications, and performance: A survey. *Knowledge-Based Systems, 151*, 78–94.
25. Goyal, P., Chhetri, S. R., & Canedo, A. (2019). *dyngraph2vec: Capturing network dynamics using dynamic graph representation learning*. Preprint. arXiv:1809.02657.
26. Ribeiro, L. F. R., Saverese, P. H. P., & Figueiredo, D. R. (2017). struc2vec: Learning node representations from structural identity. In *Proceedings of the 23rd ACM SIGKDD International Conference on Knowledge Discovery and Data Mining* (pp. 385–394). New York: ACM.
27. Vishwanathan, S. V. N., Schraudolph, N. N., Kondor, R., & Borgwardt, K. M. (2010). Graph kernels. *Journal of Machine Learning Research, 11*, 1201–1242.
28. Schuhmacher, M., & Ponzetto, S. P. (2014). Knowledge-based graph document modeling. In *Proceedings of the 7th ACM International Conference on Web Search and Data Mining* (pp. 543–552). New York: ACM.
29. Auer, S., Bizer, C., Kobilarov, G., Lehmann, J., Cyganiak, R., & Ives, Z. (2007). DBpedia: A nucleus for a web of open data. In *The Semantic Web* (pp. 722–735). Berlin: Springer.
30. Mikolov, T., Sutskever, I., Chen, K., Corrado, G., & Dean, J. (2013). Distributed representations of words and phrases and their compositionality. In *Proceedings of the 26th International Conference on Neural Information Processing Systems–Volume 2*, NIPS'13 (pp. 3111–3119). Red Hook: Curran Associates

31. Garcia-Gasulla, D., Ayguadé, E., Labarta, J., Béjar, J., Cortés, U., Suzumura, T., et al. (2017). A visual embedding for the unsupervised extraction of abstract semantics. *Cognitive Systems Research, 42*, 73–81.
32. Glorot, X., & Bengio, Y. (2010). Understanding the difficulty of training deep feedforward neural networks. In *Proceedings of the Thirteenth International Conference on Artificial Intelligence and Statistics* (pp. 249–256).
33. Bergstra, J., & Bengio, Y. (2012). Random search for hyper-parameter optimization. *Journal of Machine Learning Research, 13*, 281–305.
34. Kingma, D. P., & Ba, J. (2014). *Adam: A method for stochastic optimization*. Preprint. arXiv:1412.6980.
35. Pu, Y., Gan, Z., Henao, R., Yuan, X., Li, C., Stevens, A., & Carin, L. (2016). Variational autoencoder for deep learning of images, labels and captions. In *Advances in neural information processing systems* (pp. 2352–2360). Red Hook: Curran Associates
36. Goodfellow, I., Pouget-Abadie, J., Mirza, M., Xu, B., Warde-Farley, D., Ozair, S., et al. (2014). Generative adversarial nets. *Advances in Neural Information Processing Systems*, 2672–2680.
37. Narayanan, A., Chandramohan, M., Venkatesan, R., Chen, L., Liu, Y., & Jaiswal, S. (2017). graph2vec: Learning distributed representations of graphs. Preprint arXiv:1707.05005.

Chapter 10
Dynamic Graph Embedding

10.1 Introduction

Understanding and analyzing graphs is an essential topic that has been widely studied over the past decades. Many real-world problems can be formulated as link predictions in graphs [1–4]. For example, link prediction in an author collaboration network [1] can be used to predict potential future author collaboration. Similarly, new connections between proteins can be discovered using protein interaction networks [5], and new friendships can be predicted using social networks [6]. Recent work on obtaining such predictions uses graph representation learning. These methods represent each node in the network with a fixed dimensional embedding and map link prediction in the network space to the nearest neighbor search in the embedding space [7]. It has been shown that such techniques can outperform traditional link prediction methods on graphs [8, 9].

Existing works on graph representation learning primarily focus on static graphs of two types: (1) aggregated, consisting of all edges until time T; and (2) snapshot, which comprise of edges at the current time step t. These models learn latent representations of the static graph and use them to predict missing links [8–14]. However, real networks often have complex dynamics which govern their evolution.

In this chapter, we aim to capture the underlying network dynamics of evolution. Given temporal snapshots of graphs, our goal is to learn a representation of nodes at each time step while capturing the dynamics such that we can predict their future connections. Learning such representations is a challenging task. First, the temporal patterns may exist over varying period lengths. Second, different vertices may have different patterns. Capturing such variations is extremely challenging. Existing research builds upon simplified assumptions to overcome these challenges. Methods including DynamicTriad [15], DynGEM [16], and TIMERS [17] assume that the patterns are of short duration (length 2) and only consider the previous

© Springer Nature Switzerland AG 2020
S. R. Chhetri, M. A. Al Faruque, *Data-Driven Modeling
of Cyber-Physical Systems Using Side-Channel Analysis*,
https://doi.org/10.1007/978-3-030-37962-9_10

time step graph to predict new links. Furthermore, DynGEM and TIMERS make the assumption that the changes are smooth and use regularization to disallow rapid changes.

10.1.1 Research Challenges

In this chapter, we present a model which overcomes the above challenges. *dyngraph2vec* uses multiple non-linear layers to learn structural patterns in each network. Furthermore, it uses recurrent layers to learn the temporal transitions in the network. The lookback parameter in the recurrent layers controls the length of temporal patterns learned. We focus our experiments on the task of link prediction. We compare dyngraph2vec with the state-of-the-art algorithms for dynamic graph embedding and show its performance on several real-world networks including collaboration networks and social networks. Our experiments show that using a deep model with recurrent layers can capture temporal dynamics of the networks and significantly outperform the state-of-the-art methods on link prediction. We emphasize that the work in this chapter is targeted towards link prediction and not node classification.

10.1.2 Contribution

Overall, the contributions made by the work [18] presented in this chapter are as follows:

1. We propose *dyngraph2vec*, a dynamic graph embedding model which captures temporal dynamics.
2. We demonstrate that capturing network dynamics can significantly improve the performance on link prediction.
3. We present variations of our model to show the key advantages and differences.
4. We publish a library, DynamicGEM,[1] implementing the variations of our model and state-of-the-art dynamic embedding approaches.

10.2 Related Work

Graph representation learning techniques can be broadly divided into two categories: (1) static graph embedding, which represents each node in the graph with a single vector; and (2) dynamic graph embedding, which considers multiple

---

[1] https://github.com/palash1992/DynamicGEM.

snapshots of a graph and obtains a time series of vectors for each node. Most analysis has been done on static graph embedding. Recently, however, some works have been devoted to studying dynamic graph embedding.

10.2.1   Static Graph Embedding

Methods to represent nodes of a graph typically aim to preserve certain properties of the original graph in the embedding space. Based on this observation, methods can be divided into (1) distance preserving and (2) structure preserving. Distance preserving methods devise objective functions such that the distance between nodes in the original graph and the embedding space has similar rankings. For example, Laplacian eigenmaps [19] minimize the sum of the distance between the embeddings of neighboring nodes under the constraints of translational invariance, thus keeping the nodes close in the embedding space. Similarly, graph factorization [10] approximates the edge weight with the dot product of the nodes' embeddings, thus preserving distance in the inner product space. Recent methods have gone further to preserve higher order distances. Higher order proximity embedding (HOPE) [9] uses multiple higher order functions to compute a similarity matrix from a graph's adjacency matrix and uses singular value decomposition (SVD) to learn the representation. GraRep [12] considers the node transition matrix and its higher powers to construct a similarity matrix.

On the other hand, structure-preserving methods aim to preserve the roles of individual nodes in the graph. *node2vec* [8] uses a combination of breadth-first search and depth-first search to find nodes similar to a node in terms of distance and role. Recently, deep learning methods to learn network representations have been proposed. These methods inherently preserve the higher order graph properties including distance and structure. SDNE [20], DNGR [21], and VGAE [22] use deep auto-encoders for this purpose. Some other recent approaches use graph convolutional networks to learn inherent graph structure [23–25].

10.2.2   Dynamic Graph Embedding

Embedding dynamic graphs is an emerging topic still under investigation. Some methods have been proposed to extend static graph embedding approaches by adding regularization [17, 26]. DynGEM [27] uses the learned embedding from previous time step graphs to initialize the current time step embedding. Although it does not explicitly use regularization, such initialization implicitly keeps the new embedding close to the previous. DynamicTriad [15] relaxes the temporal smoothness assumption but only considers patterns spanning two-time steps. TIMERS [17] incrementally updates the embedding using incremental singular value decomposition (SVD) and reruns SVD when the error increases above a threshold.

DYLINK2VEC [28] learns embedding of links (node pairs instead of nodes) and uses temporal functions to learn patterns over time.

Link embedding renders the method non-scalable for graphs with high density. Our model uses recurrent layers to learn temporal patterns over long sequences of graphs and multiple fully connected layers to capture intricate patterns at each time step.

10.2.3 Dynamic Link Prediction

Several methods have been proposed on dynamic link prediction without emphasis on graph embedding. Many of these methods use probabilistic non-parametric approaches [29, 30]. NonParam [29] uses kernel functions to model the dynamics of individual node features influenced by the neighbor features. Another class of algorithms uses matrix and tensor factorization to model link dynamics [31, 32]. Further, many dynamic link prediction models have been proposed for specific applications including recommendation systems [33] and attributed graphs [34]. These methods often have assumptions about the inherent structure of the network and require node attributes. Our model, however, extends the traditional graph embedding framework to capture network dynamics.

10.3   Motivating Example

We consider a toy example to motivate the idea of capturing network dynamics. Consider an evolution of graph G, $\mathcal{G} = \{G_1, \ldots, G_T\}$, where G_t represents the state of graph at time t. The initial graph G_1 is generated using the stochastic block model [35] with 2 communities (represented by colors indigo and yellow in Fig. 10.1), each with 500 nodes. To properly demonstrate the changes in the community, we have only shown a total of 50 nodes (25 from each of the community) and shown only two migrating nodes in Fig. 10.1. The in-block and cross-block probabilities are set to 0.1 and 0.01, respectively. The evolution pattern can be defined as a three-step process. In the first step (shown in Fig. 10.1a), we randomly and uniformly select 10 nodes (colored red in Fig. 10.1) from the yellow community. In step two (shown in Fig. 10.1b), we randomly add 30 edges between each of the selected nodes in step one and random nodes in the Indigo community. This is similar to having more than cross-block probability but less than in-block probability. In step three (shown in Fig. 10.1c), the community membership of the nodes selected in step 2 is changed from yellow to indigo. Similarly, the edges (colored red in Fig. 10.1) are either removed or added to reflect the cross-block and in-block connection probabilities. Then, for the next time step (shown in Fig. 10.1d), the same three steps are repeated to evolve the graph. Informally, this can be

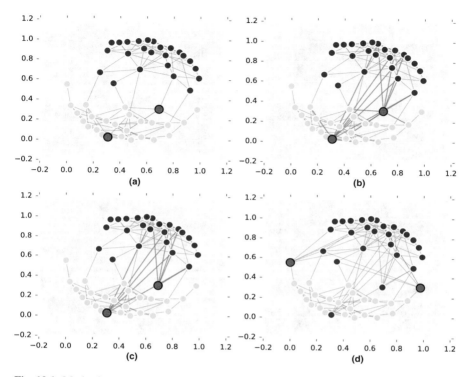

Fig. 10.1 Motivating example of network evolution—community shift

interpreted as a two-step movement of users from one community to another by initially increasing friends in the other community and subsequently moving to it.

Our task is to learn the embeddings predictive of the change in the community of the 10 nodes. Figure 10.2 shows the results of the state-of-the-art dynamic graph embedding techniques (*DynGEM*, *optimalSVD*, and *DynamicTriad*) and the three variations of our model: *dyngraph2vecAE*, *dyngraph2vecRNN*, and *dyngraph2vecAERNN* (see Methodology section for the description of the methods). Figure 10.2 shows the embeddings of nodes after the first step of evolution. The nodes selected for community shift are colored in red. We show the results for 4 runs of the model to ensure robustness. Figure 10.2a shows that DynGEM brings the red nodes closer to the edge of the yellow community but does not move any of the nodes to the other community.

Similarly, DynamicTriad results in Fig. 10.2c show that it only shifts 1–4 nodes to its actual community in the next step. The optimalSVD method in Fig. 10.2b is not able to shift any nodes. However, our *dyngraph2vecAE* and *dyngraph2vecRNN*, and *dyngraph2vecAERNN* (shown in Fig. 10.2d–f) successfully capture the dynamics and move the embedding of most of the 10 selected nodes to the indigo community, keeping the rest of the nodes intact. This shows that capturing dynamics is critical in understanding the evolution of networks.

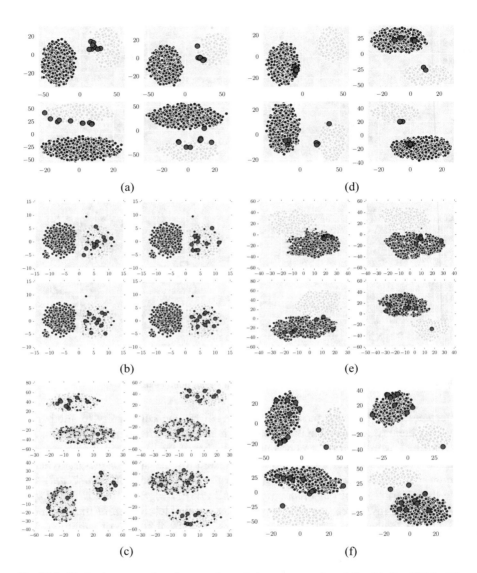

Fig. 10.2 Motivating example of network evolution—community shift. (**a**) DynGEM. (**b**) optimalSVD. (**c**) DynamicTriad. (**d**) dyngraph2vecAE. (**e**) dyngraph2vecAERNN. (**f**) dyngraph2vecRNN

10.4 Methodology

In this section, we define the problem statement. We then explain multiple variations of deep learning models capable of capturing temporal patterns in dynamic graphs. Finally, we design the loss functions and optimization approach.

10.4.1 Problem Statement

Consider a weighted graph $G(V, E)$, with V and E as the set of vertices and edges, respectively. We denote the adjacency matrix of G by A, i.e., for an edge $(i, j) \in E$, A_{ij} denotes its weight, else $A_{ij} = 0$. An evolution of graph G is denoted as $\mathcal{G} = \{G_1, \ldots, G_T\}$, where G_t represents the state of graph at time t.

We define our problem as follows: *Given an evolution of graph G, $\mathcal{G}$, we aim to represent each node v in a series of low-dimensional vector space $y_{v_1}, \ldots y_{v_t}$, where y_{v_t} is the embedding of node v at time t, by learning mappings $f_t : \{V_1, \ldots, V_t, E_1, \ldots E_t\} \rightarrow \mathbb{R}^d$ and $y_{v_i} = f_i(v_1, \ldots, v_i, E_1, \ldots E_i)$ such that y_{v_i} can capture temporal patterns required to predict $y_{v_{i+1}}$.* In other words, the embedding function at each time step uses information from graph evolution to capture network dynamics and can thus predict links with higher precision.

10.4.2 dyngraph2vec Algorithm

Our *dyngraph2vec* is a deep learning model that takes as input a set of previous graphs and generates as output the graph at the next time step, thus capturing highly non-linear interactions between vertices at each time step and across multiple time steps. Since the embedding values capture the temporal evolution of the links, it allows us to predict the next time step graph link. The model learns the network embedding at time step t by optimizing the following loss function:

$$
\begin{aligned}
L_{t+l} &= \|(\hat{A}_{t+l+1} - A_{t+l+1}) \odot \mathcal{B}\|_F^2, \\
&= \|(f(A_t, \ldots, A_{t+l}) - A_{t+l+1}) \odot \mathcal{B}\|_F^2.
\end{aligned}
\tag{10.1}
$$

Here we penalize the incorrect reconstruction of edges at time $t + l + 1$ by using the embedding at time step $t + l$. Minimizing this loss function enforces the parameters to be tuned such that it can capture evolving patterns relations between nodes to predict the edges at a future time step. The embedding at time step $t + d$ is a function of the graphs at time steps $t, t + 1, \ldots, t + l$, where l is the temporal lookback. We use a weighting matrix $\mathcal{B}$ to weight the reconstruction of observed edges higher than unobserved links as traditionally used in the literature [20]. Here, $\mathcal{B}_{ij} = \beta$ for $(i, j) \in E_{t+l+1}$, else 1, where β is a hyper-parameter controlling the weight of penalizing observed edges. Note that $\odot$ represents the elementwise product.

Algorithm 1: dyngraph2vec

Input: Graphs $\mathcal{G} = \{G_1, \ldots, G_T\}$, Dimension d, Lookback lb
Output: Embedding Vector Y
1 Generate adjacency matrices $\mathcal{A}$ from $\mathcal{G}$;
2 $\vartheta \leftarrow$ RandomInit();
3 Set $\mathcal{F} = \{(A_{t_u})\}$ for each $u \in V$, for each $t \in \{1..t\}$;
4 **for** $iter = 1 \ldots I$ **do**
5 $\quad$ $M \leftarrow$ getArchitectureInput($\mathcal{F}, lb$);
6 $\quad$ Choose L based on the architecture used;
7 $\quad$ $grad \leftarrow \partial L / \partial \vartheta$;
8 $\quad$ $\vartheta \leftarrow$ UpdateGradient($\vartheta, grad$);
9 **end**
10 $Y \leftarrow$ EncoderForwardPass(G, ϑ);
11 return Y

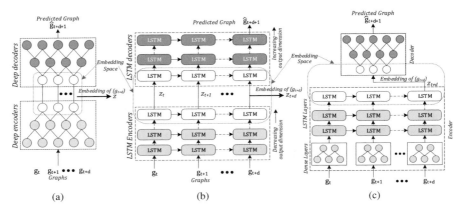

(a)   (b)   (c)

Fig. 10.3 dyngraph2vec architecture variations for dynamic graph embedding. (**a**) Dynamic graph to vector auto-encoder (*dyngraph2vecAE*). (**b**) Dynamic graph to vector recurrent neural network (*dyngraph2vecRNN*). (**c**) Dynamic graph to vector auto-encoder recurrent neural network (*dyngraph2vecAERNN*)

We propose three variations of our model based on the architecture of deep learning models as shown in Fig. 10.3: (1) *dyngraph2vecAE*, (2) *dyngraph2vecRNN*, and (3) *dyngraph2vecAERNN*. Our three methods differ in the formulation of the function $f(.)$.

A simple way to extend the auto-encoders traditionally used to embed static graphs [20] to temporal graphs is to add the information about previous l graphs as input to the auto-encoder. This model (*dyngraph2vecAE*) thus uses multiple fully connected layers to model the interconnection of nodes within and across time. Concretely, for a node u with neighborhood vector set $u_{1..t} = [a_{u_t}, \ldots, a_{u_{t+l}}]$, the hidden representation of the first layer is learned as:

$$y_{u_t}^{(1)} = f_a\big(W_{AE}^{(1)} u_{1..t} + b^{(1)}\big), \tag{10.2}$$

where f_a is the activation function, $W_{AE}^{(1)} \in \mathbb{R}^{d^{(1)} \times nl}$ and $d^{(1)}$, n and l are the dimensions of representation learned by the first layer, number of nodes in the graph, and lookback, respectively. The representation of the kth layer is defined as:

$$y_{u_t}^{(k)} = f_a\big(W_{AE}^{(k)} y_{u_t}^{(k-1)} + b^{(k)}\big). \tag{10.3}$$

Note that *dyngraph2vecAE* has $O(nld^{(1)})$ parameters. As most real-world graphs are sparse, learning the parameters can be challenging.

To reduce the number of model parameters and achieve a more efficient temporal learning, we propose *dyngraph2vecRNN* and *dyngraph2vecAERNN*. In *dyngraph2vecRNN* we use sparsely connected long short-term memory (LSTM) networks to learn the embedding. LSTM is a type of recurrent neural network (RNN) capable of handling long-term dependency problems. In dynamic graphs, there can be long-term dependencies which may not be captured by fully connected auto-encoders. The hidden state representation of a single LSTM network is defined as:

$$y_{u_t}^{(1)} = o_{u_t}^{(1)} * \tanh\big(C_{u_t}^{(1)}\big) \tag{10.4a}$$

$$o_{u_t}^{(1)} = \sigma_{u_t}\big(W_{RNN}^{(1)}\big[y_{u_{t-1}}^{(1)}, u_{1..t}\big] + b_o^{(1)}\big) \tag{10.4b}$$

$$C_{u_t}^{(1)} = f_{u_t}^{(1)} * C_{u_{t-1}}^{(1)} + i_{u_t}^{(1)} * \tilde{C}_{u_t}^{(1)} \tag{10.4c}$$

$$\tilde{C}_{u_t}^{(1)} = \tanh\big(W_C^{(1)}.\big[y_{u_{t-1}}^{(1)}, u_{1..t} + b_c^{(1)}\big]\big) \tag{10.4d}$$

$$i_{u_t}^{(1)} = \sigma\big(W_i^{(1)}.\big[y_{u_{t-1}}^{(1)}, u_{1..t}\big] + b_i^{(1)}\big) \tag{10.4e}$$

$$f_{u_t}^{(1)} = \sigma\big(W_f^{(1)}.\big[y_{u_{t-1}}^{(1)}, u_{1..t} + b_f^{(1)}\big]\big) \tag{10.4f}$$

where C_{u_t} represents the cell states of LSTM, f_{u_t} is the value to trigger the forget gate, o_{u_t} is the value to trigger the output gate, i_{u_t} represents the value to trigger the update gate of the LSTM, $\tilde{C}_{u_t}$ represents the new estimated candidate state, and b represents the biases. There can be l LSTM networks connected in the first layer, where the cell states and hidden representation are passed in a chain from $t - l$ to t LSTM networks. The representation of the kth layer is then given as follows:

$$y_{u_t}^{(k)} = o_{u_t}^{(k)} * \tanh\big(C_{u_t}^{(k)}\big) \tag{10.5a}$$

$$o_{u_t}^{(k)} = \sigma_{u_t}\big(W_{RNN}^{(k)}\big[y_{u_{t-1}}^{(k)}, y_{u_t}^{(k-1)}\big] + b_o^{(k)}\big) \tag{10.5b}$$

The problem with passing the sparse neighborhood vector $u_{1..t} = [a_{u_t}, \ldots, a_{u_{t+l}}]$ of node u to the LSTM network is that the LSTM model parameters (such as the number of memory cells, number of input units, output units, etc.) needed to learn a low dimension representation become large. Rather, the LSTM network may be able to better learn the temporal representation if the sparse neighborhood vector is reduced to a low dimension representation. To achieve this, we propose a variation of *dyngraph2vec* model called *dyngraph2vecAERNN*. In *dyngraph2vecAERNN* instead of passing the sparse neighborhood vector, we use a fully connected encoder to initially acquire low-dimensional hidden representation given as follows:

$$y_{u_t}^{(p)} = f_a\left(W_{AERNN}^{(p)} y_{u_t}^{(p-1)} + b^{(p)}\right) \tag{10.6}$$

where p represents the output layer of the fully connected encoder. This representation is then passed to the LSTM networks.

$$y_{u_t}^{(p+1)} = o_{u_t}^{(p+1)} * \tanh\left(C_{u_t}^{(p+1)}\right) \tag{10.7a}$$

$$o_{u_t}^{(p+1)} = \sigma_{u_t}\left(W_{AERNN}^{(p+1)}\left[y_{u_{t-1}}^{(p+1)}, y_{u_t}^{(p)}\right] + b_o^{(p+1)}\right) \tag{10.7b}$$

Then the hidden representation generated by the LSTM network is passed to a fully connected decoder.

10.4.3   Optimization

We optimize the loss function defined above to get the optimal model parameters. By applying the gradient with respect to the decoder weights on Eq. 10.1, we get:

$$\frac{\partial L_t}{\partial W_*^{(K)}} = \left[2(\hat{A}_{t+1} - A_{t+1}) \odot \mathcal{B}\right]\left[\frac{\partial f_a\left(Y^{(K-1)}W_*^{(K)} + b^{(K)}\right)}{\partial W_*^{(K)}}\right],$$

where $W_*^{(K)}$ is the weight matrix of the penultimate layer for all the three models. For each individual model, we back propagate the gradients based on the neural units to get the derivatives for all previous layers. For the LSTM based *dyngraph2vec* models, back propagation through time is performed to update the weights of the LSTM networks.

After obtaining the derivatives, we optimize the model using stochastic gradient descent (SGD) [36] with adaptive moment estimation (Adam)[37]. The algorithm is specified in Algorithm 1.

10.5   Experiments

In this section, we describe the datasets used and establish the baselines for comparison. Furthermore, we define the evaluation metrics for our experiments and parameter settings. All the experiments were performed on a 64 bit Ubuntu 16.04.1

LTS system with Intel (R) Core (TM) i9-7900X CPU with 19 processors, 10 CPU cores, 3.30 GHz CPU clock frequency, 64 GB RAM, and two Nvidia Titan X, each with 12 GB memory.

10.5.1 Datasets

We conduct experiments on two real-world datasets and a synthetic dataset to evaluate our proposed algorithm. We assume that the proposed models are aware of all the nodes, and that no new nodes are introduced in subsequent time steps. Rather, the links between the existing nodes change with a certain temporal pattern. The datasets are summarized in Table 10.1.

Stochastic Block Model (SBM)—Community Diminishing In order to test the performance of various static and dynamic graph embedding algorithms, we generated synthetic SBM data with two communities and a total of 1000 nodes. The cross-block connectivity probability is 0.01 and in-block connectivity probability is set to 0.1. One of the communities is continuously diminished by migrating the 10–20 nodes to the other community. A total of 10 dynamic graphs are generated for the evaluation. Since SBM is a synthetic dataset, there is no notion of time steps.

Hep-th [1] The first real-world dataset used to test the dynamic graph embedding algorithms is the collaboration graph of authors in High Energy Physics Theory conference. The original dataset contains abstracts of papers in High Energy Physics Theory conference in the period from January 1993 to April 2003. Hence, the resolution of the time step is 1 month. For our evaluation, we consider the last 50 snapshots of this dataset. From this dataset 2000 nodes are sampled for training and testing the proposed models.

Autonomous Systems (AS) [38] The second real-world dataset utilized is a communication network of who-talks-to-whom from the BGP (Border Gateway Protocol) logs. The dataset contains 733 instances spanning from November 8, 1997 to January 2, 2000. Hence, the resolution of the time step for AS dataset is 1 month as well. For our evaluation, we consider a subset of this dataset which contains the last 50 snapshots. From this dataset 2000 nodes are sampled for training and testing the proposed models.

Table 10.1 Dataset statistics

Name	SBM	Hep-th	AS
Nodes n	1000	150–14,446	7716
Edges m	56,016	268–48,274	487–26,467
Time steps T	10	136	733

10.5.2   Baselines

We compare our model with the following state-of-the-art static and dynamic graph embedding methods:

- *Optimal Singular Value Decomposition* (**OptimalSVD**) [39]: It uses the singular value decomposition of the adjacency matrix or its variation (i.e., the transition matrix) to represent the individual nodes in the graph. The low rank SVD decomposition with largest d singular values is then used for graph structure matching, clustering, etc.
- *Incremental Singular Value Decomposition* (**IncSVD**) [40]: It utilizes a perturbation matrix which captures the changing dynamics of the graphs and performs additive modification on the SVD.
- *Rerun Singular Value Decomposition* (**RerunSVD** or TIMERS) [17]: It utilizes the incremental SVD to get the dynamic graph embedding; however, it also uses a tolerance threshold to restart the optimal SVD calculation when the incremental graph embedding starts to deviate.
- *Dynamic Embedding using Dynamic Triad Closure Process* (**dynamicTriad**) [15]: It utilizes the triadic closure process to generate a graph embedding that preserves structural and evolution patterns of the graph.
- *Deep Embedding Method for Dynamic Graphs* (**dynGEM**) [16]: It utilizes deep auto-encoders to incrementally generate embedding of a dynamic graph at snapshot t by using only the snapshot at time $t - 1$.

10.5.3   Evaluation Metrics

In our experiments, we evaluate our model on link prediction at time step $t + 1$ by using all graphs until the time step t. We use mean average precision (MAP) as our metrics. $precision@k$ is the fraction of correct predictions in the top k predictions. It is defined as $P@k = \frac{|E_{pred}(k) \cap E_{gt}|}{k}$, where E_{pred} and E_{gt} are the predicted and ground truth edges, respectively. MAP averages the precision over all nodes. It can be written as $\frac{\sum_i AP(i)}{|V|}$ where $AP(i) = \frac{\sum_k precision@k(i) \cdot \mathbb{I}\{E_{pred_i}(k) \in E_{gt_i}\}}{|\{k : E_{pred_i}(k) \in E_{gt_i}\}|}$ and $precision@k(i) = \frac{|E_{pred_i}(1:k) \cap E_{gt_i}|}{k}$. $P@k$ values are used to test the top predictions made by the model. MAP values are more robust and average the predictions for all nodes. High MAP values imply that the model can make good predictions for most nodes.

10.6   Results and Analysis

In this section, we present the performance result of various models for link prediction on different datasets. We train the model on graphs from time step t to $t + l$ where l is the lookback of the model, and predict the links of the graph at time step $t + l + 1$. The lookback l is a model hyper-parameter. For an evolving graph with T steps, we perform the above prediction from $T/2$ to T and report the average MAP of link prediction. Furthermore, we also present the performance of models when an increasing length of the graph sequence is provided in the training data. Unless explicitly mentioned, for the models consisting of a recurrent neural network, a lookback value of 3 is used for the training and testing purpose.

10.6.1   SBM Dataset

The MAP values for various algorithms with SBM dataset with a diminishing community is shown in Fig. 10.4. The MAP values shown are for link prediction with embedding sizes *64*, *128*, and *256*. This figure shows that our methods *dyngraph2vecAE*, *dyngraph2vecRNN*, and *dyngraph2vecAERNN* all have higher MAP values compared to the rest of the baselines except for *dynGEM*. The *dynGEM* algorithm is able to have higher MAP values than all the algorithms. This is due to the fact that *dynGEM* also generates the embedding of the graph at snapshot $t + 1$ using the graph at snapshot t. Since in our SBM dataset the node-migration criteria are introduced only a one-time step earlier, the *dynGEM* node embedding technique is able to capture these dynamics. However, the proposed *dyngraph2vec* methods also achieve average MAP values within $\pm 1.5\%$ of the MAP values achieved by *dynGEM*. Notice that the MAP values of SVD based methods increase as the embedding size increases. However, this is not the case for *dynTriad*.

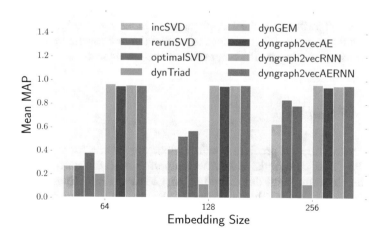

Fig. 10.4 MAP values for the SBM dataset

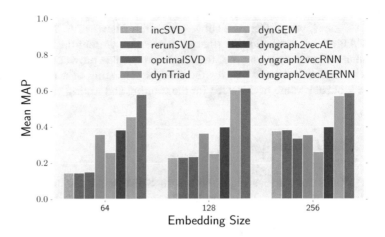

Fig. 10.5 MAP values for the Hep-th dataset

10.6.2   Hep-th Dataset

The link prediction results for the Hep-th dataset are shown in Fig. 10.5. The proposed *dyngraph2vec* algorithms outperform all the other state-of-the-art static and dynamic algorithms. Among the proposed algorithms, *dyngraph2vecAERNN* has the highest MAP values, followed by *dyngraph2vecRNN* and *dyngraph2vecAE*, respectively. The *dynamicTriad* is able to perform better than the SVD based algorithms. Notice that *dynGEM* is not able to have higher MAP values than the *dyngraph2vec* algorithms in the Hep-th dataset. Since dyngraph2vec utilizes not only $t - 1$ but also $t - l - 1$ time steps to predict the link for the time step t, it has higher performance compared to other state-of-the-art algorithms.

10.6.3   AS Dataset

The MAP value for link prediction with various algorithms for the AS dataset is shown in Fig. 10.6. *dyngraph2vecAERNN* outperforms all the state-of-the-art algorithms. The algorithm with the second highest MAP score is *dyngraph2vecRNN*. However, *dyngraph2vecAE* has a higher MAP only with a lower embedding of size 64. SVD methods are able to improve their MAP values by increasing the embedding size. However, they are not able to outperform the *dyngraph2vec* algorithms.

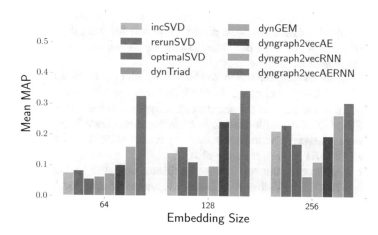

Fig. 10.6 MAP values for the AS dataset

Table 10.2 Average MAP values over different embedding sizes

Method	Average MAP		
	SBM	Hep-th	AS
IncrementalSVD	0.4421	0.2518	0.1452
rerunSVD	0.5474	0.2541	0.1607
optimalSVD	0.5831	0.2419	0.1152
dynamicTriad	0.1509	0.3606	0.0677
dynGEM	**0.9648**	0.2587	0.0975
dyngraph2vecAE (lb=3)	0.9500	0.3951	0.1825
dyngraph2vecAE (lb=5)	–	**0.512**	**0.2800**
dyngraph2vecRNN (lb=3)	**0.9567**	0.5451	0.2350
dyngraph2vecRNN	–	**0.7290** (lb = 8)	**0.313** (lb = 10)
dyngraph2vecAERNN (lb=3)	**0.9581**	0.5952	0.3274
dyngraph2vecAERNN	–	**0.739 (lb = 8)**	**0.3801 (lb = 10)**

lb Lookback value

10.6.4 MAP Exploration

The summary of MAP values for different embedding sizes (64, 128, and 256) for different datasets is presented in Table 10.2. The top three highest MAP values are highlighted in bold. For the synthetic SBM dataset, the top three algorithms with highest MAP values are *dynGEM*, *dyngraph2VecAERNN*, and *dyngraph2vecRNN*, respectively. Since the changing pattern for the SBM dataset is introduced only at time step $t - 1$, *dynGEM* is able to better predict the links. The model architecture of *dynGEM* and *dyngraph2vecAE* is only different on what data are fed to train the model. In *dyngraph2vecAE*, we essentially feed more data depending on the size of the lookback. The lookback size increases the model complexity. Since the SBM dataset does not have temporal patterns evolving for more than one-time steps, the dyngraph2vec models are only able to achieve comparable but not better result compared to *dynGEM*.

For the Hep-th dataset, the top three algorithms with the highest MAP values are *dyngraph2VecAERNN*, *dyngraph2VecRNN*, and *dyngraph2VecAE*, respectively. In fact, compared to the state-of-the-art algorithm *dynamicTriad*, the proposed models *dyngraph2VecAERNN*(with lookback = 8), *dyngraph2VecRNN* (with lookback = 8), and *dyngraph2VecAE*(with lookback = 5) obtain ≈105%, ≈102%, and ≈42% higher average MAP values, respectively.

For the AS dataset, the top three algorithms with the highest MAP values are *dyngraph2VecAERNN*, *dyngraph2VecRNN*, and *dyngraph2VecAE*, respectively. Compared to the state-of-the-art *rerunSVD* algorithm, the proposed models *dyngraph2VecAERNN*(with lookback = 10), *dyngraph2VecRNN* (with lookback = 10), and *dyngraph2VecAE* (with lookback = 5) obtain ≈137%, ≈95%, and ≈74% higher average MAP values, respectively.

These results show that the dyngraph2vec variants are able to capture the graph dynamics much better than most of the state-of-the-art algorithms in general.

10.6.5   Hyper-Parameter Sensitivity: Lookback

One of the important parameters for time series analysis is how much in the past the method looks to predict the future. To analyze the effect of lookback on the MAP score we have trained the *dyngraph2Vec* algorithms with various lookback values. The embedding dimension is fixed to 128. The lookback size is varied from 1 to 10. We then tested the change in MAP values with the real word datasets AS and Hep-th.

Performance of *dyngraph2Vec* algorithms with various lookback values for the Hep-th dataset is presented in Fig. 10.7. It can be noticed that increasing lookback values consistently increase the average MAP values. Moreover, it is interesting to notice that *dyngraph2VecAE* although has increased in performance until lookback

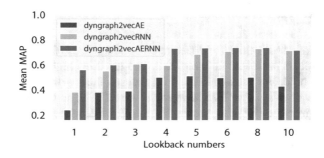

Fig. 10.7 Mean MAP values for various lookback numbers for Hep-th dataset

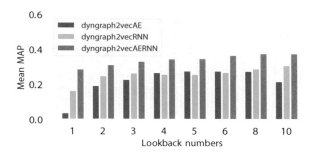

Fig. 10.8 Mean MAP values for various lookback numbers for AS dataset

size of 8, its performance is decreased for lookback value of 10. Since it does not have memory units to store the temporal patterns like the recurrent variations, it relies solely on the fully connected dense layers to encode to the pattern. This seems rather ineffective compared to the *dyngraph2VecRNN* and *dyngraph2vecAERNN* for the Hep-th dataset. The highest MAP values achieved by *dyngraph2vecAERNN* is 0.739 for the lookback size of 8.

Similarly, the performance of the proposed models while changing the lookback size for AS dataset is presented in Fig. 10.8. The average MAP values also increase with the increasing lookback size in the AS dataset. The highest MAP value of 0.3801 is again achieved by *dyngraph2vecAERNN* with the lookback size of 10. The *dyngraph2vecAE* model, initially, has comparable and sometimes even higher MAP value with respect to *dyngraph2vecRNN*. However, it can be noticed that for the lookback size of 10, the *dyngraph2vecRNN* outperforms *dyngraph2vecAE* model consisting of just the fully connected neural networks. In fact, the MAP value does not increase after the lookback size of 5 for *dyngraph2vecAE*.

10.6.6 Length of Training Sequence Versus MAP Value

In this section, we present the impact of length of graph sequence supplied to the models during training on its performance. In order to conduct this experiment, the graph sequence provided as training data is increased one step at a time. Hence, we use graph sequence of length 1 to $t \in [T, T + 1, T + 2, T + 3, \ldots, T + n]$ to predict the links for graph at time step $t \in [T + 1, T + 2, \ldots, T + n + 1]$, where $T \geq lookback$. The experiment is performed on Hep-th and AS dataset with a fixed lookback size of 8. The total sequence of data is 50 and it is split between 25 for training and 25 for testing. Hence, in the experiment, the training data sequence increases from a total of 25 sequences to 49 graph sequence. The results in Figs. 10.9 and 10.10 show the average MAP values for predicting the links starting the graph sequence at 26th to all the way to 50th time step, where each time step represents a month.

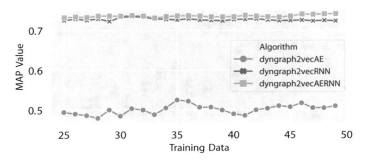

Fig. 10.9 MAP value with increasing amount of temporal graphs added in the training data for Hep-th dataset (lookback = 8)

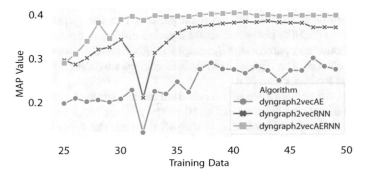

Fig. 10.10 MAP value with increasing amount of temporal graphs added in the training data for AS dataset (lookback = 8)

The result of increasing the amount of graph sequence in training data for Hep-th dataset is shown in Fig. 10.9. It can be noticed that for both the *RNN* and *AERNN* the increasing amount of graph sequence in the data does not drastically increase the MAP value. For *dyngraph2vecAE* there is a slight increase in the MAP value towards the end.

On the other hand, increasing the amount of graph sequence for the AS dataset during training gives a gradual improvement in link prediction performance in the testing phase. However, they start converging eventually after seeing 80% (total of 40 graph sequence) of the sequence data.

10.7   Discussion

Model Variation It can be observed that among different proposed models, the recurrent variation was capable of achieving higher average MAP values. These architectures are efficient in learning short- and long-term temporal patterns and provide an edge in learning the temporal evolution of the graphs compared to the fully connected neural networks without recurrent units.

Dataset We observe that depending on the dataset, the same model architecture provides different performance. Due to the nature of data, it may have different temporal patterns, periodic, semi-periodic, stationary, etc. Hence, to capture all these patterns, we found out that the models have to be tuned specifically to the dataset.

Sampling One of the weakness of the proposed algorithms is that the model size (in terms of the number of weights to be trained) increases based on the size of the nodes considered during the training phase. To overcome this, the nodes have been sampled. Currently, we utilize uniform sampling of the nodes to mitigate this issue. However, we believe that a better sampling scheme that is aware of the graph properties may further improve its performance.

Large Lookbacks While it is desirable to test large lookback values for learning the temporal evolution with the current hardware resources, we constantly ran into resource exhausted error with lookbacks greater than 10.

10.8 Summary

In this chapter, we introduced dyngraph2vec, a dynamic data-driven model for capturing temporal patterns in dynamic networks. It learns the evolution patterns of individual nodes and provides an embedding capable of predicting future links with higher precision. We propose three variations of our model based on the architecture with varying capabilities. The experiments show that the proposed model can capture temporal patterns on synthetic and real datasets and outperform state-of-the-art methods in link prediction. There are several directions for future work: (1) interpretability by extending the model to provide more insight into network dynamics and better understand temporal dynamics; (2) automatic hyper-parameter optimization for higher accuracy; and (3) graph convolutions to learn from node attributes and reduce the number of parameters. The unsupervised data-driven approach can be applied to abstract temporally evolving non-euclidean data present in a cyber-physical system.

References

1. Gehrke, J., Ginsparg, P., & Kleinberg, J. (2003). Overview of the 2003 KDD cup. *ACM SIGKDD Explorations, 5*(2), 149–151.
2. Freeman, L. C. (2000). Visualizing social networks. *Journal of Social Structure, 1*(1), 4.
3. Theocharidis, A., Van Dongen, S., Enright, A., & Freeman, T. (2009). Network visualization and analysis of gene expression data using BioLayout Express 3D. *Nature Protocols, 4*, 1535–1550.

4. Goyal, P., Sapienza, A., Ferrara, E. (2018). Recommending teammates with deep neural networks. In *Proceedings of the 29th on Hypertext and Social Media* (pp. 57–61). New York: ACM.
5. Pavlopoulos, G. A., Wegener, A.-L., & Schneider, R. (2008). A survey of visualization tools for biological network analysis. *Biodata Mining, 1*(1), 12.
6. Wasserman, S., & Faust, K. (1994). *Social network analysis: Methods and applications* (Vol. 8). Cambridge: Cambridge University Press.
7. Goyal, P., & Ferrara, E. (2018). Graph embedding techniques, applications, and performance: A survey. *Knowledge-Based Systems, 151*, 78–94.
8. Grover, A., & Leskovec, J. (2016). Node2vec: Scalable feature learning for networks. In *Proceedings of the 22nd International Conference on Knowledge Discovery and Data Mining* (pp. 855–864). New York: ACM.
9. Ou, M., Cui, P., Pei, J., Zhang, Z., & Zhu, W. (2016). Asymmetric transitivity preserving graph embedding. In *Proceedings of the 22nd ACM SIGKDD* (pp. 1105–1114).
10. Ahmed, A., Shervashidze, N., Narayanamurthy, S., Josifovski, V., & Smola, A. J. (2013). Distributed large-scale natural graph factorization. In *Proceedings of the 22nd International Conference on World Wide Web* (pp. 37–48). New York: ACM.
11. Perozzi, B., Al-Rfou, R., & Skiena, S. (2014). DeepWalk: Online learning of social representations. In *Proceedings 20th International Conference on Knowledge Discovery and Data Mining* (pp. 701–710).
12. Cao, S., Lu, W., & Xu, Q. (2015). GraRep: Learning graph representations with global structural information. In *KDD15* (pp. 891–900).
13. Tang, J., Qu, M., Wang, M., Zhang, M., Yan, J., & Mei, Q. (2015). Line: Large-scale information network embedding. In *Proceedings 24th International Conference on World Wide Web* (pp. 1067–1077).
14. Goyal, P., Hosseinmardi, H., Ferrara, E., & Galstyan, A. (2018). Embedding networks with edge attributes. In *Proceedings of the 29th on Hypertext and Social Media* (pp. 38–42). New York: ACM.
15. Zhou, L., Yang, Y., Ren, X., Wu, F., & Zhuang, Y. (2018). Dynamic network embedding by modelling triadic closure process. In *Thirty-Second AAAI Conference on Artificial Intelligence*.
16. Goyal, P., Kamra, N., He, X., & Liu, Y. (2018). *DynGEM: Deep embedding method for dynamic graphs*. Preprint. arXiv:1805.11273.
17. Zhang, Z., Cui, P., Pei, J., Wang, X., & Zhu, W. (2017). *TIMERS: Error-bounded SVD restart on dynamic networks*. Preprint. arXiv:1711.09541.
18. Goyal, P., Chhetri, S. R., & Canedo, A. (2019). *dyngraph2vec: Capturing network dynamics using dynamic graph representation learning*. Preprint. arXiv:1809.02657.
19. Belkin, M., & Niyogi, P. (2001). Laplacian eigenmaps and spectral techniques for embedding and clustering. In *NIPS'01 Proceedings of the 14th International Conference on Neural Information Processing Systems: Natural and Synthetic* (Vol. 14, pp. 585–591).
20. Wang, D., Cui, P., & Zhu, W. (2016). Structural deep network embedding. In *Proceedings of the 22nd International Conference on Knowledge Discovery and Data Mining* (pp. 1225–1234). New York: ACM.
21. Cao, S., Lu, W., & Xu, Q. (2016). Deep neural networks for learning graph representations. In *Proceedings of the Thirtieth AAAI Conference on Artificial Intelligence* (pp. 1145–1152). Palo Alto: AAAI Press.
22. Kipf, T. N., Welling, M. (2016). *Variational graph auto-encoders*. Preprint. arXiv:1611.07308.
23. Kipf, T. N., & Welling, M. (2016). *Semi-supervised classification with graph convolutional networks*. Preprint. arXiv:1609.02907.
24. Bruna, J., Zaremba, W., Szlam, A., & LeCun, Y. (2013). *Spectral networks and locally connected networks on graphs*. Preprint. arXiv:1312.6203.
25. Henaff, M., Bruna, J., & LeCun, Y. (2015). *Deep convolutional networks on graph-structured data*. Preprint. arXiv:1506.05163.

26. Zhu, L., Guo, D., Yin, J., Ver Steeg, G., & Galstyan, A. (2016). Scalable temporal latent space inference for link prediction in dynamic social networks. *IEEE Transactions on Knowledge and Data Engineering, 28*(10), 2765–2777.
27. Goyal, P., Kamra, N., He, X., & Liu, Y. (2017). DynGEM: Deep embedding method for dynamic graphs. In *IJCAI International Workshop on Representation Learning for Graphs*.
28. Rahman, M., Saha, T. K., Hasan, M. A., Xu, K. S., & Reddy, C. K. (2018). *DyLink2Vec: Effective feature representation for link prediction in dynamic networks.* Preprint. arXiv:1804.05755.
29. Sarkar, P., Chakrabarti, D., & Jordan, M. (2012). *Nonparametric link prediction in dynamic networks.* Preprint. arXiv:1206.6394.
30. Yang, S., Khot, T., Kersting, K., & Natarajan, S. (2016). Learning continuous-time Bayesian networks in relational domains: A non-parametric approach. In *Thirtieth AAAI Conference on Artificial Intelligence*.
31. Dunlavy, D. M., Kolda, T. G., & Acar, E. (2011). Temporal link prediction using matrix and tensor factorizations. *ACM Transactions on Knowledge Discovery from Data (TKDD), 5*(2), 10.
32. Ma, X., Sun, P., & Wang, Y. (2018). Graph regularized nonnegative matrix factorization for temporal link prediction in dynamic networks. *Physica A: Statistical Mechanics and Its Applications, 496*, 121–136.
33. Talasu, N., Jonnalagadda, A., Pillai, S. S. A., & Rahul, J. (2017). A link prediction based approach for recommendation systems. In *2017 International Conference on Advances in Computing, Communications and Informatics (ICACCI)* (pp. 2059–2062). Piscataway: IEEE.
34. Li, J., Cheng, K., Wu, L., & Liu, H. (2018). Streaming link prediction on dynamic attributed networks. In *Proceedings of the Eleventh ACM International Conference on Web Search and Data Mining* (pp. 369–377). New York: ACM.
35. Wang, Y. J., & Wong, G. Y. (1987). Stochastic blockmodels for directed graphs. *Journal of the American Statistical Association, 82*(397), 8–19.
36. Rumelhart, D. E., Hinton, G. E., & Williams, R. J. (1988). In *Neurocomputing: Foundations of research*. J. A. Anderson & E. Rosenfeld (Eds.) (pp. 696–699). New York: ACM.
37. Kingma, D. P., & Ba, J. (2014). *Adam: A method for stochastic optimization.* Preprint. arXiv:1412.6980.
38. Leskovec, J., Kleinberg, J., & Faloutsos, C. (2005). Graphs over time: Densification laws, shrinking diameters and possible explanations. In *Proceedings of the Eleventh ACM SIGKDD International Conference on Knowledge Discovery in Data Mining* (pp. 177–187). New York: ACM.
39. Ou, M., Cui, P., Pei, J., Zhang, Z., & Zhu, W. (2016). Asymmetric transitivity preserving graph embedding. In *Proceedings of the 22nd ACM SIGKDD International Conference on Knowledge Discovery and Data Mining* (pp. 1105–1114). New York: ACM.
40. Brand, M. (2006). Fast low-rank modifications of the thin singular value decomposition. *Linear algebra and its applications, 415*(1), 20–30.

Index

A

Acoustic emission
  electromagnetic source, 16–17
  equation of motion, 14–15
  mechanical source, 17–18
  natural rotor oscillation frequency, 15
  stator natural frequency, 16
  system description, 14
Acoustic leakage analysis
  leakage exploitation, 20–21
  leakage quantification, 20
  side-channel leakage model, 18–20
Acoustic sensor, 39, 53, 120, 170, 176
AdaBoost algorithm, 31
Architecture Analysis and Design Language
  (AADL), 2
Artificial intelligence algorithms, 186
Attack model
  success rate reconstruction, 85–86
  test objects, with reconstruction, 86–87
Automatic model generation method, 117
Autonomous systems (AS), 219

C

Classification model, 26–27
Clustering algorithm
  $DT_{product}$, 164–165
  performance of, 171
Compiler alteration, 46
Computer-aided design (CAD), 185, 198–199
Computer-aided manufacturing (CAM)
  software, 12, 185

Conditional generative adversarial model
  (CGAN)
  acoustic sensor, 120
  automatic model generation method, 117
  of CPPS, 115, 116
  energy and signal flows, 114, 116
  experimental data collection, 121
  graph generation, 120
  model generation, 116–119
  multiple sub-systems, 115
  security analysis results, 122–125
  storage for security analysis, 119
  system-level methodology, 116
  3D printer, 118, 120
  training results, 121–122
Confidentiality, 43, 89
Cyber-physical additive manufacturing system,
  13
Cyber-physical production systems (CPPS),
  111
  conditional generative adversarial
    network-based model, 114
  G_{CPPS}, 119–120
Cyber-physical systems (CPS)
  computational components, 1
  CPS modeling languages such, 2
  cross-domain attacks, 4
  cross-domain components, 1
  cross-layer components, 1
  formal and provable deterministic
    properties, 2
  heterogeneous and hybrid components, 1
  non-euclidean data, 5–6

© Springer Nature Switzerland AG 2020
S. R. Chhetri, M. A. Al Faruque, *Data-Driven Modeling
of Cyber-Physical Systems Using Side-Channel Analysis*,
https://doi.org/10.1007/978-3-030-37962-9

Printed in the United States
By Bookmasters